T0239017

SpringerBriefs in Applied Sciences and Technology

SpringerBriefs present concise summaries of cutting-edge research and practical applications across a wide spectrum of fields. Featuring compact volumes of 50 to 125 pages, the series covers a range of content from professional to academic.

Typical publications can be:

- A timely report of state-of-the art methods
- An introduction to or a manual for the application of mathematical or computer techniques
- A bridge between new research results, as published in journal articles
- A snapshot of a hot or emerging topic
- An in-depth case study
- A presentation of core concepts that students must understand in order to make independent contributions

SpringerBriefs are characterized by fast, global electronic dissemination, standard publishing contracts, standardized manuscript preparation and formatting guidelines, and expedited production schedules.

On the one hand, **SpringerBriefs in Applied Sciences and Technology** are devoted to the publication of fundamentals and applications within the different classical engineering disciplines as well as in interdisciplinary fields that recently emerged between these areas. On the other hand, as the boundary separating fundamental research and applied technology is more and more dissolving, this series is particularly open to trans-disciplinary topics between fundamental science and engineering.

Indexed by EI-Compendex, SCOPUS and Springerlink.

Rachid Ababou · Juliette Chastanet ·
Jean-Marie Côme · Manuel Marcoux ·
Michel Quintard

Uncertainty Analyses in Environmental Sciences and Hydrogeology

Methods and Applications to Subsurface Contamination

 Springer

Rachid Ababou
Institut de Mécanique des Fluides de
Toulouse (IMFT)
Toulouse, France

Jean-Marie Côme
GINGER
Lyon, France

Michel Quintard
Institut de Mécanique des Fluides de
Toulouse (IMFT)
Toulouse, France

Juliette Chastanet
GINGER
Lyon, France

Manuel Marcoux
Institut de Mécanique des Fluides de
Toulouse (IMFT)
Toulouse, France

ISSN 2191-530X   ISSN 2191-5318 (electronic)
SpringerBriefs in Applied Sciences and Technology
ISBN 978-981-99-6240-2   ISBN 978-981-99-6241-9 (eBook)
https://doi.org/10.1007/978-981-99-6241-9

This Springer imprint is published by the registered company Springer Nature Singapore Pte Ltd.
The registered company address is: 152 Beach Road, #21-01/04 Gateway East, Singapore 189721, Singapore

Paper in this product is recyclable.

Acknowledgments

This book is based in part on previous research work performed in the framework of the ESPER project, funded by ADEME, the French environmental protection agency (Grant award number 1672C0022). The authors wish to acknowledge contributions from graduate students and junior researchers at the Institute of Fluid Mechanics in Toulouse: R. Leroux (IMFT post-doc) on fuzzy and possibilistic approaches; Y. Soumane (IMFT master student 2018-2019) on the 3D analytical model and ESPER-1 probabilistic package. The first author also thanks Pr. K. Alastal (IMFT visiting scientist 2019–2020) for additional help on MATLAB implementation of the 3D analytical model, and Pr. N. El Moçayd (UM6P) for his useful 2022 comment on Sobol indices versus Polynomial Chaos.

Contents

Chapter 1
Introduction, Objectives

The different alternatives for managing a contaminated site are based on a cost/benefit balance that requires predictive modeling to assess mass depletion and duration of contamination source in the subsurface. For instance, the ESPER[1] project (*Evaluation and Sensitivity of models for Predicting the depletion and Remediation of organic contamination sources in subsurface*) was aimed at developing a methodology and a software tool for facilitating the incorporation of uncertainty and sensitivity analyses into such subsurface contamination models.

Let us first clarify from the outset the term "uncertainty propagation."

Mathematical methods of uncertainty analyses can be viewed as "uncertainty propagation" methods, in the sense that they can "propagate" (or "carry") the uncertainty from the input parameters to the outputs of the model. The model may be analytical: a simple example is the first order decay model $C(t) = C_0 \exp(-\lambda t)$. Or it can be a more complex quasi-analytical model containing special functions and simple integrals. Or else, it can be a fully discretized space–time numerical model such as the MODFLOW- SURFACT™ code. It can also be, possibly, a polynomial "meta-model" ("surrogate model") derived from the initial model. The metamodel is essentially an input/output response function, derived from the original model, and which can be used to simplify the uncertainty propagation process (as will be explained later).

This book reviews and illustrates various approaches and methods for uncertainty analysis, and presents various models (from simple to complex) to illustrate these uncertainty analyses, including some model-specific results, and also, site-specific results, from the ESPER project in particular.

This book is organized as follows. After the present introduction, the next chapter (Chap. 2) covers different methods for carrying out uncertainty analysis, starting with an overview of different approaches and concepts. Chapter 3 develops a comprehen-

---

[1] The ESPER project was funded in part by ADEME, the French environmental protection agency.

R. Ababou et al., *Uncertainty Analyses in Environmental Sciences and Hydrogeology*, SpringerBriefs in Applied Sciences and Technology, https://doi.org/10.1007/978-981-99-6241-9_1

sive setting for probabilistic uncertainty quantification with random input parameters (including multivariate vectors of parameters). Chapter 4 focuses the review on fuzzy variables, comparing probabilistic vs. fuzzy approaches to uncertainty, and also, presents combined random/fuzzy approaches based on *possibility theory* (extension of fuzzy variables theory). The rest of the book is devoted to subsurface contamination models, with a view to their implementation with uncertain parameters. Thus, we consider in Chap. 5 a variety of equational models of contaminant transport, some analytical, others semi-analytical with simple integrals or special functions, and yet other fully numerical models (space–time discretized). Finally, in Chap. 6, we develop several examples of uncertainty analyses using some of the previous models with uncertain parameters. Various techniques and approaches are used (fuzzy or probabilistic, Monte-Carlo simulations with or without a meta-model). They are applied to a fully 3D semi-analytical model of solute advection–dispersion from a dissolving source (this model is used as a test case for our probabilistic *ESPER-1* Monte Carlo uncertainty package), and finally, to a complex 3D computer code (MODFLOW- SURFACT™) tested on a real contaminated site with uncertain parameters (based on response function metamodeling).

Chapter 2
Overview of Uncertainty Propagation Methods

This section presents, first, a summarized overview of approaches and concepts, before going into specific methods later in more detail (probabilistic and fuzzy approaches to uncertainty in hydrogeology, Monte-Carlo procedures, Metamoding, etc.). Let us first quote some general reference texts in the literature. See [1] on probability laws and random variables as well as random processes; [2] on random spatial fields; [3] on probability and stochastic modeling for risk analysis. Concerning hydrogeologic flows and contaminant transport underground, see [4–6]. Many more works will be cited along the way when discussing and reviewing topics of uncertainty analysis and contaminant transport.

2.1 Approaches to Uncertainty

Before tackling the different methods of uncertainty propagation, it should be first recognized that there are several ways to characterize and quantify uncertainty (i.e., the uncertain parameters and variables). Several approaches, summarized below, will be considered:

- The uncertain input parameters (and the output variables or criteria) are all considered to be random variables or random vectors following known probability laws; or else…
- The uncertain input parameters (and the output variables or criteria) are all considered to be fuzzy variables possessing known membership functions, or likelihood functions (cf. fuzzy set theory of [7, 8]), or else…
- The uncertain input parameters follow a combination of probability laws and fuzzy membership functions, or "possibility functions" (extension of fuzzy theory such as the "Hybrid approach" and the Independent Random Sets IRS approach, as will be seen).

R. Ababou et al., *Uncertainty Analyses in Environmental Sciences and Hydrogeology*, SpringerBriefs in Applied Sciences and Technology, https://doi.org/10.1007/978-981-99-6241-9_2

2.2 Monte Carlo Simulations (with or Without a Metamodel)

2.2.1 Direct Monte Carlo Simulations

Essentially two broad classes of procedures can be considered in order to "propagate" uncertainty:

- Monte Carlo simulations of outputs from multiple replicates of uncertain input parameters
- Analytical characterization of outputs uncertainty (outputs probability law and/ or moments).

The most usual method to propagate uncertainty from inputs to outputs is by running the model for multiple replicates of the inputs: this is known as "Monte Carlo simulations." This principle of repeated simulations can be applied to different approaches of uncertainty (probabilistic, fuzzy, or both), and to any kind of model (analytical or numerical).

However, for some simple models, Monte Carlo simulations may not be required at all. A fully analytical uncertainty propagation may be possible. In such cases, it is possible to express explicitly the probability law of the model outputs given the probability law of the inputs.

In addition, approximate analytical methods of uncertainty propagation can also be implemented without recourse to Monte Carlo simulations (e.g., using first order Taylor development around the mean). However, these approximations usually neglect the nonlinearity of the model, and they are limited to moderate or small uncertainty.

To sum up, Monte Carlo simulations are required in many cases, especially with complex models. The objective is to perform "M" repeated simulations of the model for "M" different replicates of the uncertain inputs, in order to analyze the distribution of the uncertain outputs. One of the crucial questions is then: *How Many Monte Carlo Simulations Are Needed ?* ($M = 100?M = 1000?M = 10\,000?$). A brief answer is that Monte Carlo simulations converge slowly, with statistical error proportional to $\sim 1/\sqrt{M}$. More precisely, *statistical theory* results indicate that, when repeatedly sampling from a population, the precision on the sample mean $\widehat{m}_X = \left(\sum_{m=1}^{m=M} X^{(m)}\right)/M$ (with respect to the true unknown mean m_X) is proportional to $1/\sqrt{M}$ where M is sample size (number of replicates drawn from the population). This $1/\sqrt{M}$ behavior can be used as a guide toward answering the question.

2.2.2 Indirect Monte Carlo Simulations via a Metamodel

Principle of the Metamodel Approach

As will be seen in more detail later on, given an input/output model (such as Darcy flow and advection–diffusion transport), the multiple replicate/Monte Carlo approach to uncertainty propagation can be applied in two different ways, either directly to the model itself (as suggested earlier), or else indirectly, by applying Monte Carlo simulations to a "metamodel." The metamodel (to be constructed) should provide a "response function" that is less costly to compute than the model itself. The goal of the response function is to provide a risk criterion (an output of the model) that is cheaply calculated as a simple function of the input parameters.

Metamodel construction techniques (building a response function)

Quantifying uncertainty via a metamodel implies the preliminary construction of a response function. The response function is defined in a P-dimensional space, where P is the number of uncertain parameters of the model.

Response functions have been constructed using a variety of techniques. Some techniques are quite basic, like empirical nonlinear regression based on low-degree polynomials, applicable in practice to a relatively small or moderate number of uncertain parameters P.

Other metamodel construction techniques are more sophisticated, like Polynomial Chaos (see [13] and references therein), or like Kriging estimation implemented in parameter space (rather than the usual spatial kriging in 1,2,3-D Euclidian space). Concerning kriging as a tool for metamodeling in parameter space, see for instance the DICE R-packages presented by [9]. In this specialized field, current efforts are geared toward techniques that might be efficient for high-dimensional parameter spaces. Thus, [10] developed and tested a stochastic kriging technique (with added random noise) using "Tensor Markov" kernels. The kernel is essentially the prior covariance matrix of generalized Gauss-Markov processes in the space of the uncertain parameters. Ding and Zhang [10] tested their algorithms for high dimensional parameter space with hundreds and up to 10 000 uncertain parameters.

Both Polynomial Chaos expansions, and Kriging Estimation (viewed as "Gaussian process regression"), are used to construct metamodels within the OpenTURNS software (see the section on "*Metamodels*" in [11]. A remarkable fact with Polynomial Chaos (PC) metamodels is that, once a PC metamodel has been constructed, Sobol indices used for sensitivity analyses can be computed analytically from the PC coefficients (see details in [12]).

2.2.3 Summary on Direct Versus Indirect Monte Carlo Simulations

In summary, while the "direct" Monte Carlo approach performs multiple simulations by directly implementing the full input/output model for a large number of replicates of input parameters, the "indirect" Monte Carlo approach calculates a large number of replicates of the prescribed output criterion via repeated application of a cheaper "metamodel," usually with polynomial response function constructed in advance for a prescribed criterion. Multiple replicates of the output criterion are computed cheaply from the metamodel response function. However, the construction of the response function requires running the full model multiple times at least once for the given output criterion. A disadvantage of this "indirect" approach is that the metamodel "response function" must be constructed for each desired output criterion.

For more details on theory and algorithms with metamodels, see previously mentioned references. For instance, see [13] on polynomial chaos metamodels, and [11] on running metamodels and other methods with the *OpenTURNS* platform. Concerning the use of kriging as "Gaussian process regression" in uncertainty analyses, see again [11], and also [10] on using kriging with Tensor Markov kernels in high-dimensional parameter space. Examples of various uncertainty propagation methods will be presented throughout this book in the review sections below, and in the applications Chaps. 5 and 6.

2.3 Joint Sets of Parameters, Cross-Correlations, Design of Experiments

When generating multiple replicates of input parameters, the case of jointly uncertain sets of parameters should be considered. For instance, the dispersion coefficient $D(m^2/s)$ and groundwater velocity $V(m/s)$ may be partially correlated through dispersivity length scale (m). Or the permeability $k(m^2)$ and porosity Φ may be partially correlated through the Kozeny-Carman relationship. The joint set of input parameters may in fact comprise dozens of partially cross-correlated parameters, possibly together with other independent parameters. In such cases, it should be decided whether to propagate "jointly" or "one-by-one" the parameters' uncertainty (the latter approach is easier but does not convey the correlation amongst parameters). Appropriate "joint" sampling of the parameters can involve generating "equiprobable" multivariate samples, and/or other sampling techniques known as "Design of Experiments" (DoE).

2.3.1 Sampling and Uncertainty Propagation for Each Input Parameter (One-By-One)

This case is relatively easy to implement: the same uncertainty propagation procedure is implemented by repeatedly sampling each uncertain input parameter while the others are fixed to some reference value (one-by-one analysis). There remains the question of how to sample the M replicates of the single uncertain parameter.

Equiprobable sampling for one-by-one uncertainty analysis (univariate case)

Equiprobable sampling of input parameters can be recommended by default. It is such that the number of replicates in each parameter interval is proportional to its probability to belong to that interval. For instance, Fig. 2.1 illustrates the equiprobable sampling of a Gaussian parameter with zero mean and unit variance. The random parameter has the same probability of falling in each interval: therefore, intervals widths are non-uniform. In the context of Monte Carlo simulations, the sampling strategy will be to draw the same number of replicates in each equiprobable interval. This graphic was produced by us in MATLAB using random number generators and the inverse Gaussian CDF, for only a few tens of replicates (another similar graphic can be found in [13]. Note that extreme high and low values are conspicuously absent for such small samples: the sampled PDF appears truncated. ("Importance sampling" techniques can circumvent this problem by approximating the output probability law to enable targeting important extreme regions of input parameter space).

The equiprobable sampling approach can be extended to cross-correlated bivariate and multivariate cases (uniform, Gaussian, Log-normal, or other). The equiprobability criterion can be enforced in multivariate cases via the *LHS* sampling technique (*LHS: Latin Hypercube Sampling*) to be reviewed further below. Before this, let us first see how the univariate "one-by-one" uncertainty propagation through a model can be performed and then, for instance, used for sensitivity analysis purposes.

One-by-one uncertainty propagation from inputs to outputs

Performing uncertainty analysis for a set of N uncertain parameters taken *one-by-one* (separately) provides a somewhat limited characterization of output uncertainty (for instance parameters cross-correlations are ignored), but it has a few advantages…

   i. *One-by-one* analysis is simple to implement repeatedly, in a sequential loop over the set of N parameters $(P_1, \ldots, P_n, \ldots, P_N)$.

  ii. For instance, if the first parameter P_1 is considered uncertain (the others being fixed), a set of M equiprobable replicates $\left(P_1^{(1)}, \ldots, P_1^{(m)}, \ldots, P_1^{(M)} \right)$ is generated, and correspondingly, M replicates of the output variables are obtained—such as pollutant concentration at a given space–time point, NAPL source mass at the same given time, and/or output criteria "R" inferred from output variables (like the time to reach a given concentration level…).

 iii. This is repeated for parameter $P_2, \ldots$, parameter P_N. Each execution of the model (analytical or numerical) delivers a set of replicates of the output variables, which

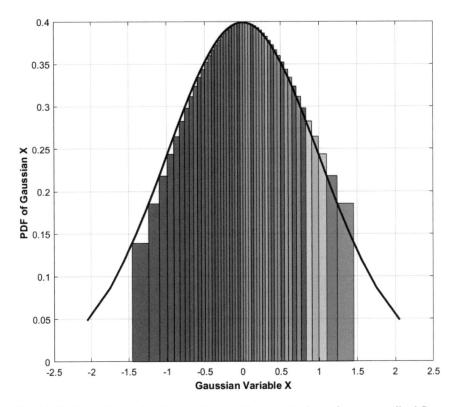

Fig. 2.1 Equiprobable sampling for a single uncertain parameter drawn from a normalized Gaussian distribution (Univariate Gaussian). This graphic representation was programmed in MATLAB; extreme low/high intervals are truncated due to the relatively small number of replicates in this sample

         are uncertain. The outer loop ($n = 1, ..., N$) over uncertain input parameters taken one-by-one is parallelizable. Furthermore, the inner Monte Carlo loop over replicates ($m = 1, ..., M$) is parallelizable too.

iv. *One-by-one* analysis can be used to compare the variability of a given output criterion "$R(P_1, P_2, \ldots, P_N)$" with respect to several different parameters taken *one-by-one* (the others remaining fixed): thus $R(P_1)$ is analyzed first, then $R(P_2)$, etc...., where $R(P_1)$ is defined as $R(P_1) \equiv R\big(P_1, P_2^0, \ldots, P_N^0\big)$, etc.... Note that the fixed parameters labeled P_j^0 are considered "certain"; they are frozen parameters, to be used as reference values.

Uncertainty propagation and sensitivity coefficients (one-by-one)

The above one-by-one technique can be viewed as a way to implement *sensitivity analysis* via uncertainty propagation. For example, the results might show that the Output Criterion R is more uncertain (has a larger variance) with respect to Parameter 1 than with respect to Parameter 2.

In the context of groundwater contamination, criterion R could be the time taken by the peak of a concentration plume, or by a given threshold concentration isovalue, to reach a given distance. It could also be the source "lifetime" (e.g., the time taken by NAPL source to decrease to 90% of its initial mass).

Example: $P1$ = Permeability; $P2$ = Porosity; Output Criterion R = 90% *Source* Lifetime.

The sensitivity coefficients "S" of criterion R with respect to $P1$ and $P2$ are the partial derivatives:

$$S_{P1}(R) \equiv \partial R / \partial P_1 \quad ; \quad S_{P2}(R) \equiv \partial R / \partial P_2$$

Another analytical way to perform sensitivity analysis is to compute the contribution of each uncertain parameter to the global uncertainty of the output. This can be accomplished analytically for simple algebraic models of the form Output $= f$ (InputParameters) using Taylor expansion around mean values, or Mean Value First Order method (MVFO). See for instance the sensitivity analyses of [14] for corrosion pits growth $Z(t)$ on a steel nuclear waste canister. These authors used the MVFO method to calculate directly the contribution of each uncertain parameter, P_n, to the total variance of the output $Z(t)$, and ranked them accordingly: a large variance contribution of P_n implies that the model is very sensitive to that parameter P_n. We present later in Sect. 5.3. the corrosion pit problem in nuclear waste canisters with a simple geochemical model.

Other methods can be used for *joint* (instead of "one-by-one") sensitivity analysis: see the brief presentation of Sobol indices and their relation to Polynomial Chaos coefficients, in the next section below (Sect. 2.3.2.).

2.3.2 Joint Uncertainty Propagation for a Parameter Set $\{P_1, P_2, ..., P_K\}$

We examine here techniques for jointly sampling the set of uncertain parameters, and then propagating uncertainty through a model, for a set of K input parameters $\{P_1, P_2, \ldots, P_K\}$ with $K > 1$, where the different parameters may or may not be cross-correlated.

First let us briefly reconsider the univariate case ($K = 1$). The principle of equiprobable sampling was illustrated earlier in this section for a single uncertain parameter: see previous Fig. 2.1, where iso-probability intervals of a single Gaussian variable were shown. Note that equiprobable sampling does not need to be random in this example: one can choose a set of N deterministically distributed values (N samples) of the uncertain parameter to be sampled. Another possibility is dividing the variable axis into K deterministic "boxes," and taking $N = L \times K$ samples with $L \gg 1$. Then the same number of samples L can be drawn from each of the iso-probability boxes of Fig. 2.1, possibly at random within each box. To sum up, the sampling schemes just discussed could be considered as basic examples of what is known as "Design of Experiment" (DoE). However, this "DoE" topic is more relevant for *multiple* input parameters ("factors"). The *Latin Hypercube Sampling* scheme, introduced below, is a parsimonious sampling scheme for the multivariate case.

Design of Experiments (DoE) for 2 factors: Latin Hypercube Sampling

The term "Design of Experiment" (DoE) refers to the design of a sampling scheme, especially where several variables or "factors" are involved (probabilistic or not). If the variables can be considered probabilistic, the previous "equiprobable" principle could be applied again. However, as the number of variables increases, the number of samples will increase too. In order to circumvent this problem, a parsimonious sampling known as *Latin Hypercube Sampling* (LHS) is frequently used. Let us briefly describe the LHS scheme for the bivariate case of two random parameters (X, Y) (*correlated or not*):

- First, partition the (X, Y) plane into iso-probability boxes according to the joint probability distribution $F_{X,Y}(x, y) \equiv Pr\{X \leq x, Y \leq y\}$. Usually these 2D boxes are not of uniform size (except for a joint *uniform* distribution).
- Secondly, sample these boxes by marking them in such a way that only one box is selected in each row, and one box in each column. The point samples (X,Y) are then positioned within the marked boxes, (e.g., at box centers).

Two examples of *Latin Hypercube Sampling* are shown in the next figure (Fig. 2.2), for a bi-variate uniform distribution (Fig. 2.2 **Top**), and for a bivariate Gaussian distribution (Fig. 2.2 **Bottom**), where equiprobable regions are shown as 2D cells. Note: these graphs were generated by us from our Matlab scripts; similar schematics can also be found in [13] and other authors.

Design of Experiments (DoE) with multiple factors: LHS and other schemes

Here we revisit the DoE issue (Design of Experiments), re-examining Latin Hypercube Sampling (LHS) in comparison with other schemes, and discussing points of practical interest *(e.g., how to treat cross-correlated parameters having disparate probability laws)*.

First, here is an overview of classical references on LHS schemes in geosciences, where the parameters to be sampled are sometimes called "factors": see [15] on parameter aggregation in hydrological models, [16] on sensitivity analyses and uncertainty propagation with the hydrologic model MIKE SHE; [17] on Design of Experiments for pesticides, or [18] concerning data assimilation and uncertainty in hydrology.

Several codes for simulating hydro-geological flow-transport incorporate tools for solving inverse problems, performing sensitivity analyses, and/or propagating uncertainty from inputs to outputs. This is the case of the 3D iTOUGH2 code for multiphase flow and transport in porous media. The "i" in "iTOUGH2" stands for "inverse": solving inverse problems by optimization, but also, performing sensitivity analyses and propagating uncertainty. In iTOUGH2, multiple replicates of input parameters can be generated by Latin Hypercube Sampling (LHS) according

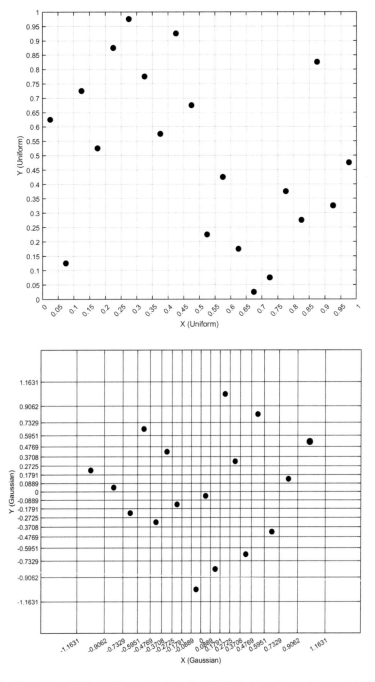

Fig. 2.2 Illustration of 2D Latin Hypercube Sampling (LHS) for bi-variate problems, with Uniform (X, Y) and Gaussian (X, Y) distributions. *Top* Equiprobable LHS sampling of the Uniform Bivariate. *Bottom* Equiprobable LHS sampling of the Gaussian Bivariate

to prescribed probability laws. These features are also related to another inverse code (PEST, iTOUGH2-PEST). See [19] for an optimization User Guide (after the PEST software), and [20] for the iTOUGH2 v7.0 Command Reference. The latter includes LHS as a sampling scheme for Monte Carlo simulations with multiple input parameters, for some probability distributions (Uniform, Gaussian, Log-normal), albeit with ambiguous description of Log-normal parameters (*to be clarified later in this book*). Concerning iTOUGH2-PEST, see also [21].

Several other softwares have incorporated Uncertainty Quantification packages (e.g., "SmartUQ" in Comsol Multiphysics). In addition, in this book (Chap. 6), two other models of 3D flow transport will be implemented for uncertainty propagation analyses: (1) a semi-analytical 3D model of source dissolution and concentration migration (developed for Monte Carlo uncertainty analyses within our ESPER-1 package), and (2) the commercial 3D MODFLOW- SURFACT™ code applied to a real site.

Design of Experiments (DoE) and LHS: special problems

Latin Hypercube Sampling (LHS) can work for probabilistic Monte Carlo sampling of a multivariate set of input parameters $\{P_1, P_2, \ldots, P_N\}$, be they independent or dependent. However, dependence between random parameters having different probability distributions may be hard to characterize. The easiest case is that of a multivariate Gaussian set of parameters, where the dependence is entirely characterized by their $N \times N$ covariance matrix $\mathbf{C_{PP}}$, or the Log-normal multivariate distribution which can be handled in different ways through the log-transform (Sect. 3.1.2). Thus, if two parameters $\{P_1, P_2\}$ follow the bivariate Gaussian distribution, their dependence is completely characterized by their covariance C_{P1P2} or by their correlation coefficient $\rho = C_{P1P2}/(\sigma_{P1}\sigma_{P2})$. For other multivariate cases, with dependence involving mixed distributions (e.g. dependent Gaussian and Uniform variables), it may be necessary to specify a complete multivariate joint probability law. The theory of *"copulas"* could help solve this parametrization problem. See *"Copula (probability theory)"* in [22]. To sum up, provided careful specification of the joint probability law, Latin Hypercube Sampling (LHS) could be adapted to generate multiple samples of non-Gaussian cross-correlated multivariate sets of random parameters.

Sampling schemes other than LHS have been devised in the literature. Thus, "Quasi Random Sequences" or "Low Discrepancy Sequences" are available as a sampling scheme in OpenTURNS (see Sects. 5.4.1 and 6.2). Yet other samplings may be needed for evaluating the output probability distribution in narrow probability regions, for reliability analyses ("failure risk," extreme values, probability tails). Several techniques known as "importance sampling," "adaptive sampling," or "stratified sampling," have been devised for such purposes. Some of these are quite complex, involving iterations and corrections during Monte Carlo simulations. See [23, 24] for applications of importance sampling in structural systems. Other techniques are simpler to implement semi-analytically, based on approximate Taylor expansion of output probability distribution. As shown later in Sect. 5.3 for a simple model of corrosion pit growth, *first order Taylor expansion* can be used to propagate uncertainty analytically through an input/output model for performing *probabilistic failure analysis.*

Finally, we emphasize that LHS and "importance sampling" methods are applicable in two manners: (1) either directly through Monte Carlo simulations of the model of interest (e.g., flow-transport code), or else, (2) indirectly via simulations of a *metamodel* ("surrogate model") which approximates the response of the original model. Various sampling strategies can be applied during the construction/calibration phase of the metamodel, and during its exploitation via Monte Carlo simulations. Picheny et al. [24] considered a class of metamodels based on generalized kriging, and devised adaptive procedures for estimating the probability of failure of a structural system through the metamodel. They constructed a Design of Experiments such that the metamodel accurately approximates the vicinity of a boundary in design space, this boundary being defined by a target value of the response function of interest (maximum stress in the system minus allowable "failure" stress).

Joint sensitivity analyses, Sobol indices, and Polynomial Chaos coefficients

Finally, joint sensitivity analyses can be performed via "*Sobol indices*." Briefly, the first Sobol index S_{P1}^{SOBOL} quantifies the proportion of output variance that can be attributed to the sole parameter P1; S_{P1P2}^{SOBOL} quantifies the joint effects of both parameters $(P1, P2)$; etc. The "total" Sobol index $S_{Pi,TOTAL}^{SOBOL}$ quantifies the total effect of parameter P_i including its interactions with all other parameters $j \neq i$. These indices have a particular relation to Polynomial Chaos (PC) metamodels: once a PC metamodel has been constructed, Sobol indices can be computed analytically from the PC coefficients (see details in [12]).

2.4 Applications to Risk Assessment in Field Pollution

Mathematical methods of uncertainty propagation allow the quantification of model prediction uncertainties, and *in fine*, they can provide the quantitative elements for risk assessment, for instance, by developing cost/benefit balance toward the best choices for contaminated site management. In this context, methods of uncertainty propagation allow evaluating the uncertainty of costs and gains of remediation, as explained below (this is an important topic, even though we will not directly model remediation operations in this book).

Uncertain costs of remediation

Uncertainty propagation through appropriate remediation models can lead to an assessment of the uncertain financial costs linked to containment methods like hydraulic barrier, reactive barrier, Monitored Natural Attenuation ("MNA"), etc. This is important, as these costs may become prohibitive if the containment methods have to be applied on a long period of time.

Uncertain gains of remediation

Similarly, the gains resulting from remediation, and their uncertainty, can also be evaluated by propagating uncertainty through specific remediation models. The gain induced by a source zone remediation solution (such as excavation, or in situ treatment) can be compared to other remediation techniques, keeping in mind that remediation techniques do not generally treat all the contamination contained in the subsurface (due to limited efficiency of these methods, and/or, due to constraints of access to the site, etc.).

To sum up, approaches based on remediation modeling with uncertain parameters will allow a quantification of the uncertainty on the lifetime of the contaminant source and on its environmental impact, and this for several remediation scenarios.

2.5 Causes and Sources of Uncertainty

Methods of uncertainty propagation are applicable to a broader range of risk assessment issues related, in particular, to contamination phenomena in subsurface hydrology and hydrogeology: soil contamination, groundwater contamination, safety of toxic and radioactive waste disposal facilities, etc. In all these cases, uncertainty is due (*in part*) to poorly known hydrogeologic parameters (such as porosity, permeability, adsorption, and dispersivity coefficients), and also, to poorly known spatial distributions of parameters (heterogeneity). However, uncertainty in the model outputs may also be due to "model errors."

Heterogeneity versus Uncertainty

It is worth noting that heterogeneity has sometimes been treated as if it were equivalent to uncertainty in the literature. Both can be treated probabilistically. However, based on random field theory and geostatistics, we argue that the two concepts (heterogeneity, uncertainty) should be distinguished.

(i)   Thus, aquifer permeability $K(x, y)$ may be represented as a random field, possibly conditioned to honor measured permeabilities at several points. Multiple heterogeneous "Monte Carlo" replicates $K(x, y)^{(m)}$ $\{m = 1, 2, \ldots, M\}$ may then be generated by Bayesian or geostatistical methods.

(ii) At any point (x_1, y_1) other than a measurement point, the permeability $K(x_1, y_1)$ is indeed uncertain, and its distribution can be analyzed across the set of replicates $m = 1, \ldots, M$.

(iii) On the other hand, any given replicate of $K(x, y)$ is spatially variable; thus, the first replicate $K(x, y)^{(1)}$ is heterogeneous, the second also, etc.

(iv) The spatial statistical structure of random field permeability $K(x, y)$ is usually not well known: it could be considered uncertain. The mean, variance, and variogram structure of $K(x, y)$ could all be considered fuzzy, as illustrated in Fig. 2.3. References [25, 26] developed a geostatistical estimation of $K(x, y)$

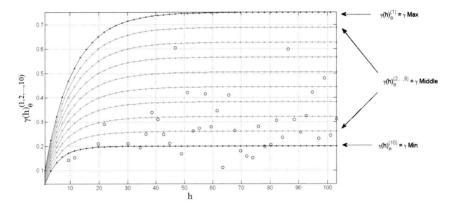

Fig. 2.3 Example of a fuzzy two-point variogram $\Gamma(h)$, where h is the distance between the two points (meters). The fuzzy variogram structure is based on two fuzzy parameters: the total variance (plateau of the variogram), and the range (or autocorrelation length). For instance here, the total variance is roughly between 0.20 and 0.75, and the range between 20 and 50 m. The blue circles correspond to calculated values of $\Gamma(h)$ from experimental data

based on kriging with a fuzzy variogram. Celmins [27] discussed the related topic of nonlinear regression and model calibration using fuzzy logic.

The above considerations indicate the subtle way in which both uncertainty and heterogeneity should be combined, ideally, in hydrogeologic field modeling.

Model errors

Finally, we have noted that "modeling errors" also contribute to uncertainty in the predicted outputs (such as contaminant concentration or flux). This type of uncertainty includes not only numerical errors, but also, conceptual model errors (e.g., incorrectly assuming Darcy's law near wells with high velocities, or incorrectly assuming spatially constant coefficients in highly heterogeneous aquifers). For a discussion on model complexity, heterogeneity, and model errors, see [28].

In summary, the causes of uncertainty are not always clearly identified. Uncertainty can be due to:

- imperfect knowledge of parameter values (measurement errors, interpolations, …);
- spatial variability of the geologic medium (imperfectly represented or ignored);
- model errors due to incorrect or forgotten mechanisms in the equational model.

Model errors, measurement errors, model validation versus refutation, the "scope" of an experiment, model entropy, and degree of freedom, are discussed in *Ababou, Sagar et al. (1992)*. In practice, it is often convenient to start with a given equational model (i.e., to accept the model), with its predefined set of input parameters (spatially constant or not), and then to translate imperfect knowledge of these parameters in terms of their "uncertainty" (randomness or fuzziness).

2.6   Uncertainty Propagation: Output "Criteria"

The remaining task is then to implement uncertainty propagation from the input parameters of the model to its outputs, or to pre-defined *criteria* (labeled "*R*") such as:

- uncertain concentration $C(L, t)$ at a given distance from the source;
- uncertain contaminant flux $\varphi(L, t)$ at a given point or a given boundary;
- uncertain time $T(L)$ taken by the contaminant to reach a given point or a given boundary located at distance L from the source.

The latter criterion, essentially, answers the question: "*what is the time taken by a given iso-concentration to reach distance L?*" It may be replaced by other temporal criteria such as: "*what is the time it takes for the NAPL source mass to reach a given percentage p% of the initial mass?*".

The main tool for uncertainty propagation will be to generate several (many) replicates of the inputs and outputs (Monte Carlo simulations), in two possible ways.

- Direct Monte Carlo: M input replicates are generated, and the model is implemented M times to generate M replicates of the outputs; or
- Indirect Monte Carlo/*metamodel* approach: a set of M_0 inputs replicates is generated, and the model is executed M_0 times to produce M_0 output replicates, leading to construction of the metamodel; the latter is then used to analyze a larger set of $M \gg M_0$ input/output replicates (in this phase, only the "cheap" metamodel is used, not the full model).

With any of these two methods, direct or indirect, the goal is to characterize the distribution of outputs or "criteria," and to perform uncertainty analysis of the contaminated site (with or without remediation plan). Note that, if input parameters are random, output criteria are also random and can be analyzed probabilistically. Risk assessment methods often rely on probabilistic concepts, like conditional probabilities, loss functions, reliability functions, mean time between failures, etc. See [29] in the context of actuarial studies, or [24] for reliability analysis in structural safety.

Thus, the time $T(L)$ taken by the contaminant to reach a given point can be analyzed probabilistically through Monte Carlo, e.g., for calculating its Cumulated Distribution Function (CDF), defined as: $F_T(\tau) = \text{Proba}\{T \leq \tau\}$. The quality of the estimated CDF is improved with an increasing number of replicates (M)…at the expense of CPU time costs. The link between uncertainty and reliability analyses is the concept of "*failure*." In subsurface contamination, a "failure" event could be defined as follows: concentration $C(\vec{x}_1, t)$ at a given point $(\vec{x}_1)$ exceeds a given target concentration C_{TARGET} (see field site application in Sect. 6.2). Such "failure" events, arising from hydrogeological space–time phenomena, are analogous to those defined in reliability analyses for industrial products like lightbulbs, or also, to dam break events in hydrology.

References

1. A. Papoulis, S.U. (Unnikrishna) Pillai, *Probability, Random Variables, and Stochastic Processes* (16 Chaps.), 4th edn. (Mc-Graw Hill, 2002), 852pp.
2. E. Vanmarcke, *Random Fields: Analysis and Synthesis* (Massachusetts Institute of Technology Press, Cambridge, 1983), p.382
3. S.E. Serrano, *Engineering Uncertainty and Risk Analysis: Balanced Approach to probability, Statistics, Stochastic Modeling, and Stochastic Differential Equations* (HydroScience Inc., Ambler, USA, 2011). ISBN 978-0-9655643-1-1, 462pp.
4. G. De Marsily, *Quantitative Hydrogeology* (Academic, New York, 1986), p.440
5. R.A. Freeze, J.A. Cherry, *Groundwater* (Prentice-Hall, USA, 1979)
6. G.F. Pinder, W.G. Gray, *Essentials of Multiphase Flow and Transport in Porous Media* (Wiley, 2008), 257p.
7. L.A. Zadeh, Fuzzy sets. Inf. Control **8**(3), 338–353 (1965)
8. L.A. Zadeh, Outline of a new approach to the analysis of complex systems and decision processes. IEEE Trans. Syst. Man Cybern. SMC-**3**, 28–44 (1973)
9. O. Roustant, D. Ginsbourger, Y. Deville, DiceKriging, DiceOptim: Two R packages for the analysis of computer experiments by kriging-based metamodeling and optimization (2012). <hal-00495766v3>
10. L. Ding, X. Zhang, Sample and computationally efficient stochastic kriging in high dimensions. Oper. Res. (2022). https://doi.org/10.1287/opre.2022.2367, https://www.researchgate.net/publication/363878439
11. M. Baudin, A. Dutfoy, B. Iooss, A.-L. Popelin, OpenTURNS: an industrial software for uncertainty quantification in simulation, in *Handbook of Uncertainty Quantification*, eds. by R. Ghanem, D. Higdon, H. Owhadi (Springer, 2017), 46p. HAL-01107849v2
12. L. Le Gratiet, S. Marelli, B. Sudret, Metamodel-based sensitivity analysis: Polynomial chaos expansions and Gaussian processes, in *Handbook of Uncertainty Quantification*, eds. by R. Ghanem, D. Higdon, H. Owhadi (2016). https://doi.org/10.48550/arXiv.1606.04273
13. S. Poles, A. Lovison, A polynomial chaos approach to robust multiobjective optimization, in *Dagstuhl Seminar Proceedings 09041: Hybrid & Robust Approaches to Multiobjective Optimization*, ed. by K. Deb et al., Dagstuhl, Germany (2009), 15p. ISSN : 1862-4405, http://drops.dagstuhl.de/opus/volltexte/2009/2000/
14. D.C.-F. Shih, G.-F. Lin, Uncertainty and importance assessment using differential analysis: an illustration of corrosion depth of spent nuclear fuel canister. Stoch. Environ. Res. Risk Assess. (SERRA) **20**, 291–295 (2006). https://doi.org/10.1007/s00477-005-0028-z
15. B. Diekkrüger, Upscaling of hydrological models by means of parameter aggregation technique, in *Dynamics of Multiscale Earth Systems*. ed. by P.H.J. Neugebauer, C. Simmer. Lecture Notes in Earth Sciences. (Springer, Berlin, 2003), pp.145–165
16. K. Christiaens, J. Feyen, Use of sensitivity and uncertainty measures in distributed hydrological modeling with an application to the MIKE SHE model. Water Resour. Res. **38**, 1169 (2002). https://doi.org/10.1029/2001WR000
17. O. Richter, B. Diekkrüger, P. Nörtersheuser, *Environmental Fate Modelling of Pesticides* (Wiley, 1996)
18. K. Beven, J. Freer, Equifinality, data assimilation, and uncertainty estimation in mechanistic modelling of complex environmental systems using the GLUE methodology. J. Hydrol. **249**, 11–29 (2001). https://doi.org/10.1016/S0022-1694(01)00421-8
19. S. Finsterle, *iTOUGH2 Universal Optimization Using the PEST Protocol: User's Guide*, August 2011. Report LBNL-3698E (revised). Earth Sciences Division, Lawrence Berkeley National Laboratory, University of California, Berkeley, CA 94720 (2011)
20. S. Finsterle, *iTOUGH2 V7.0 Command Reference*, May 2015. Report LBNL-40041 (Revised). Earth Sciences Division, Lawrence Berkeley National Laboratory, University of California, Berkeley, CA 94720 (2015)
21. S. Finsterle, Y. Zhang, Solving iTOUGH2 simulation and optimization problems using the PEST protocol. Environ. Model. Softw. **26**(2011), 959–968 (2011)

22. Wikipedia article, *Copula (probability theory)*. Web page last edited on 3 October 2021 (Consulted in November 2021) (2021), https://en.wikipedia.org/wiki/Copula_(probability_the ory)
23. R.E. Melchers, Importance sampling in structural systems. Struct. Saf. **6**(1), 3–10 (1989)
24. V. Picheny, D. Ginsbourger, O. Roustant, R.T. Haftka, N.-H. Kim, *Adaptive Designs of Experiments for Accurate Approximation of a Target Region* (2010), 13p. HAL-00319385v2, https://hal.archives-ouvertes.fr/hal-00319385v2
25. A. Bardossy, I. Bogardi, W.E. Kelly, Kriging with imprecise (fuzzy) variograms. I: theory. Math. Geol. **22**(1), 63–79 (1990)
26. A. Bardossy, I. Bogardi, W.E. Kelly, Kriging with imprecise (fuzzy) variograms. II: application. Math. Geol. **22**(1), 81–94 (1990)
27. A. Celmins, A practical approach to nonlinear fuzzy regression. SIAM J. Sci. Stat. Comput. **12**(1991), 521–546 (1991)
28. R. Ababou, B. Sagar, G. Wittmeyer, Testing procedures for spatially distributed flow models. Adv. Water Resour. **15**, 181–198 (1992)
29. A.S. Klugman, H.H. Panjer, G.E. Willmot, *Loss Models (From Data to Decisions)*. Wiley Series in Probability and Statistics, 5th edn. (Wiley & Society of Actuaries, 2019), 536pp.

Chapter 3
Review of Probabilistic Versus Fuzzy Approaches to Uncertainty Propagation in Geosciences

In this section, we develop a critical review and assessment of two types of approaches to uncertainty, based on probabilistic formulations, and based on fuzzy set theory (initiated by [9, 10]). In both approaches, the objective is to quantify and propagate uncertainty from model inputs to model outputs for risk assessment purposes. The applications targeted here are problems of subsurface contamination in hydrogeology and geosciences. As will be seen, the two types of approaches, probabilistic and fuzzy, are not necessarily mutually exclusive.

3.1 Probabilistic Characterization of Uncertainty

The textbook by [5] is a reference for some of the probabilistic concepts required in this book. However, some topics deserve particular attention here, such as multivariate Gaussian and Log-normal distributions with two or more variables, relations between moments of the Gaussian variable Y and Log-normal variable $X = \exp(Y)$, or other issues like high quantiles estimation. For this reason, this section includes a summary presentation of probability/statistics results for uncertainty analysis. We are concerned with the probabilistic characterization of random parameters, or more generally, of a set of jointly random parameters (extending the previous overview Chap. 2 and Sect. 2.3), and with related sampling/estimation issues. In what follows, uncertain parameters are considered random "variables" named "X" or "Y".

3.1.1   Univariate Probabilistic Characterization (Single Parameter)

This subsection reviews classical probabilistic concepts for characterizing a random variable, including CDF and PDF functions, quantiles, and moments, but also, questions related to their empirical estimations from samples. We will also present, at the end of Sect. 3.1.2, several algorithms to generate random variables (uniform, Gaussian, and Gaussian pairs).

PDF, CDF, and their estimation from samples

Cumulated Distribution Function (CDF). For a single random variable X, the probability of non-exceedance $Pr\{X \leq x\}$ is the so-called Cumulated Distribution Function (CDF) that characterizes completely the probability law of X. It is often denoted $F_X(x)$, where X (upper case) designates the random variable itself, while x (lower case) designates the deterministic values it can take (see also for instance the quantiles defined further below):

$$F_X(x) = \Pr\{X \leq x\} \in [0, 1] \tag{3.1}$$

Probability Density Function (PDF). If the CDF is continuous, the Probability Density Function (PDF) can be defined as the function $f_X(x) = dF_X/dx$. The density $f_X(x)$ has units of $[X^{-1}]$; thus, if X is hydraulic conductivity $K(\text{m/s})$, its PDF $f_K(k)$ has units $[(\text{m/s})^{-1}]$. On the other hand, for any infinitesimal interval $[x, x + dx]$, the quantity $dF_X(x) = f_X(x) \cdot dx$ can be interpreted as a probability increment as follows:

$$f_X(x) \cdot dx = \Pr\{x \leq X \leq x + dx\} \tag{3.2}$$

Estimation of the PDF. The PDF of X can be estimated from a sample of N replicates of the random variable X by constructing a histogram with a chosen histogram width (Δx). Note that the frequency histogram $f\%$ is dimensionless, while the PDF is a density and has units $[X^{-1}]$. Therefore, the frequency histogram $f\%$, estimated at the middle of each histogram interval $(x_{j-1/2})$, should be divided by Δx, the width of histogram intervals, to obtain the required estimate of the PDF itself $(\hat{f})$:

$$\hat{f}\left(x_{j-\frac{1}{2}}\right) = \frac{f\%\left(x_{j-1/2}\right)}{100\Delta x}\{j = 1, 2, \ldots, N\} \tag{3.3}$$

Estimation of the CDF. Two methods are available for estimating the CDF from a sample of N replicates of the random variable X: (i) cumulated histogram, or (ii) estimation by points (method of Hazen). The method by points should be preferred. It consists simply in taking the following two steps: (a) sort the data sample $\left\{X^{(j)}, j = 1, \ldots, N\right\}$ in increasing order, and (b) attribute a constant probability increment $\Delta P = 1/N$ to each new data in increasing order, as follows (*we may later omit the "sort" superscript, which indicates "sorted data"*):

$$\hat{F}\left(X_j^{\text{SORT}}\right) = \frac{j^{\text{SORT}} - 1/2}{N} \left\{ j^{\text{SORT}} = 1, 2, \ldots, N \right\} \tag{3.4}$$

Examples of PDF/CDF estimation and fit. Figures 3.1 and 3.2 show an estimated PDF histogram and a point estimation of the CDF for a synthetic dataset of size $N = 100$ sampled from a Gaussian distribution. The fitted Gaussian PDF and CDF curves are obtained by the method of moments, i.e., by inserting estimated empirical moments in the proposed theoretical law. Here the empirical moments were $m_X = 9.9273$, $\sigma_X = 0.9536$, to be compared to theoretical moments $m_X = 10$, $\sigma_X = 1$. A remarkable fact is that the empirical CDF seems better fitted to the theoretical Gaussian CDF, compared to the worse fit of the PDF histogram. These results would be even much worse if the data did not come from a Gaussian distribution.

Final remarks on empirical probability law estimation:

(i) The only consistent estimator for the PDF is the histogram with a chosen *histogram width Δx*; there is no alternative: pointwise estimation of a PDF is simply not possible (inconsistent).

(ii) The CDF can be estimated, like the PDF, using a histogram (cumulated). This is less satisfactory than the point estimator of the CDF, which has better resolution and depends only on the data (not on *histogram width Δx*). This estimator is consistent and unbiased.

(iii) The goodness of fit of the CDF to a theoretical CDF can be evaluated by the *Ki2 or Khi2* test, or by the *Kolmogorov–Smirnov* test (*not detailed here*).

Moments and their estimation from samples

The random variable X can also be characterized partially through some of its moments: (1) the mean, (2) the variance, and (normalized) higher order moments like (3) skewness, and (4) kurtosis. Estimators of these quantities are described in Papoulis & Pillai (2022). Unbiased estimators of the mean m and variance σ^2 are:

$$\hat{m} = \frac{1}{N} \sum_{j=1}^{j=N} X^{(j)}; \quad \widehat{\sigma^2} = \frac{1}{N-1} \sum_{j=i}^{j=N} \left(X^{(j)} - \hat{m}\right)^2 \tag{3.5}$$

Notice the intentional division by *N-1* instead of N for *unbiased* variance estimation. The standard deviation σ can then be estimated from the square root of $\widehat{\sigma^2}$:

$$\hat{\sigma} \approx \sqrt{\frac{1}{N-1} \sum_{j=1}^{j=N} \left(X^{(j)} - \hat{m}\right)^2} \tag{3.6}$$

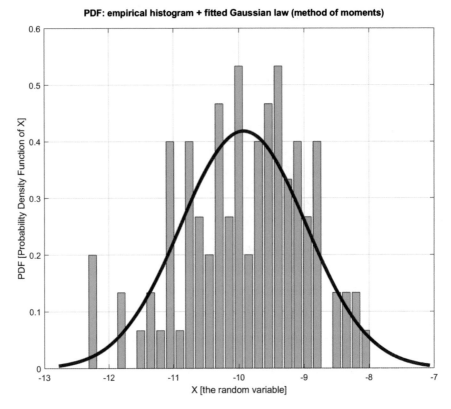

Fig. 3.1 Histogram estimation of the PDF of a sample of size $N = 100$ generated from a Gaussian distribution ($m_X = 10$, $\sigma_X = 1$), showing also the theoretical Gaussian PDF (red curve) fitted to the estimated PDF histogram by the method of moments

The standard deviation estimate $\hat{\sigma}$ just above is unfortunately biased in general, but this should not be a problem except for very small samples. Physically, $\hat{\sigma}$ has the same units as X; it is a root-mean square measure of variability of X (unlike the variance $\widehat{\sigma^2}$ which has units of X squared).

Probability distributions of interest

- Uniformly distributed variable $X : \mathcal{U}[X_L, X_U]$ between a lower bound value X_L and an upper bound value X_U; the PDF $f_X(x)$ is $1/(X_U - X_L)$ inside $[X_L, X_U]$, and zero outside. Example: porosity ($X = \Phi$) with $0 < X_U < X_L < 1$.
- Gaussian ("Normal") random variable $X : \mathcal{N}\{m_X, \sigma_X^2\}$. Its PDF is:

$$\text{Gaussian PDF}: \quad f_X(x) = \frac{1}{\sqrt{2\pi\sigma_X^2}} \exp\left\{-\frac{1}{2}\left(\frac{x - m_X}{\sigma_X}\right)^2\right\} \qquad (3.7)$$

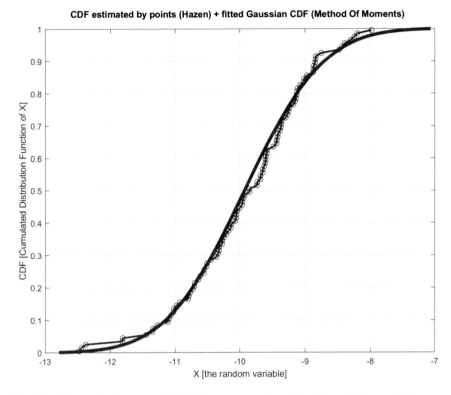

CDF estimated by points (Hazen) + fitted Gaussian CDF (Method Of Moments)

Fig. 3.2 Point estimation of the CDF of the same sample of size $N = 100$ generated from a Gaussian distribution ($m_X = 10, \sigma_X = 1$), showing also the theoretical Gaussian CDF (red curve) fitted to the estimated CDF curve by the method of moments

The Gaussian CDF can be expressed as:

$$\text{Gaussian CDF}: \ F_X(x) = \frac{1}{2}\left\{1 + \text{erf}\left(\frac{x - m_X}{\sigma_X\sqrt{2}}\right)\right\} \tag{3.8}$$

where $erf(x)$ is the classical special function called "*error function*".

- Log-Normal random variable $X \geq 0$: X is such that the Neperian logarithm of X is Gaussian, that is:

$$X : \text{LogNormal}\{m_X, \sigma_X^2\} \Leftrightarrow \text{Ln } X = Y : \mathcal{N}\{m_Y, \sigma_Y^2\}.$$

The log-normal distribution is important in applications because it allows relating positive parameters like hydraulic conductivity $K\,(\text{m/s})$ or permeability $k\,(\text{m}^2)$ to the Gaussian law...by assuming they are log-normally distributed. With this in mind, the above description of log-normal variables should be completed with a characterization of the relations between Gaussian moments $\{m_Y, \sigma_Y^2\}$ and log-normal moments $\{m_X, \sigma_X^2\}$.

The relations between moments $\{m_Y, \sigma_Y^2\}$ of the Gaussian variable Y, and moments $\{m_X, \sigma_X^2\}$ of the positive log-normal variable X, are well known (*see for instance Eq. 3.21 further below*). Some of these relations can be found in various texts, like [7], including a useful product moment formula for all moments of a Gaussian variable obtained in 1918 by Isserlis [4].

Quantiles from probabilities

Finally, it should be kept in mind that $F_X(x)$ is the dimensionless probability of non-exceedance: $F_X(x) = \text{Proba}\{X \leq x\}$. Therefore, knowledge of $F_X(x)$ leads to knowledge of quantiles "$x_{P\%}$". For instance, the 50% quantile is the *median*, and the 95% quantile ($x_{95\%}$) is the value of the uncertain parameter X which has only 5% chance of being exceeded: $F_X(x_{95\%}) = \text{Proba}\{X \leq x_{95\%}\} = 95\%$. Assuming that the probability law $F_X(x)$ is known (or estimated), the value of a quantile can be obtained by inverting the CDF $F_X(x)$. Let us give an example for a Gaussian variable. In that case, the « P%» quantile $x_{P\%}$ is obtained from:

$$x_{P\%} = m_X + \sigma_X \sqrt{2} \text{erf}^{-1}(2P - 1) \tag{3.9}$$

where erf^{-1} is the inverse error function (called *erfinv* in MATLAB). For the 95% quantile, this yields:

$$x_{95\%} = m_X + \sigma_X \sqrt{2} \, \text{erf}^{-1}(2 \times 0.95 - 1) \tag{3.10}$$

In the normalized Gaussian case, with $m_X = 0$ *and* $\sigma_X = 1$, this yields $x_{95\%} = 1.645$. Applying the same formula with $F = 0.50$ to compute the 50% quantile (median) yields $x_{50\%} = 0$ as expected.

The difficult estimation of high quantiles. The direct estimation of a quantile $X_{P\%}$ based on a finite size data sample $\{X^{(j)}, j = 1, \ldots, N\}$ has inherent limitations. The quantile $X_{P\%}$ can be estimated by direct interpolation between sample data $\{X^{(k)}, X^{(k+1)}\}$, but *only if* sample size N is larger than a critical sample size:

$$N > N_{\text{CRIT}}(P\%) = 0.5/(1 - P\%/100), \tag{3.11}$$

...or equivalently, *only if* the quantile probability (P%) is no higher than a critical probability value:

$$P\% < P\%_{\text{CRIT}}(N) = 100(1 - 0.5/N), \tag{3.12}$$

...where *the factor 100* is due to *percentages*. For a "high" quantile ($P\% > 100(1 - 0.5/N)$) it is *impossible* to estimate $X_{P\%}$ directly. In that case, $X_{P\%}$ can only be estimated indirectly as follows: (i) estimate the CDF function $\widehat{F}_X(x)$, (ii) fit a theoretical model $F_X^{\text{FIT}}(x)$ to $\widehat{F_X}(x)$, and (iii) interpolate or extrapolate $F_X^{\text{FIT}}(x)$ to obtain finally an indirect estimate of the quantile $X_{P\%}$. This limitation regarding "high quantiles" should be kept in mind in the context of uncertainty analysis. Possibly, resampling methods related to Jackkife or Bootstrap could be used to overcome this problem (but this is not treated here). Let us now give a simple *example* related to subsurface contaminant migration modeling with uncertain parameters.

Example: high quantile of concentration. Consider the problem of estimating the concentration quantile $C99.90\%$ corresponding to uncertain concentration $C(\vec{x}, t)$ calculated at a given space–time point, from "M" Monte Carlo simulations of the contaminant migration model, corresponding to "M" replicates of the uncertain model inputs. From the previous considerations on critical sample size, we have $N_{\text{CRIT}} = 0.5/(1 - P\%/100)$, and taking $P\% = 99.90\%$, we find that we need $M > 500$ Monte Carlo simulations of the model in order to estimate the concentration quantile $C99.90\%$.

3.1.2 Multivariate Probabilistic Characterization (Two or More Uncertain Parameters)

The previous concepts are here briefly extended to multivariate collections of random parameters, for example, uncertain probabilistic parameters that might be statistically dependent (cross-correlated). We also present, at the end of this section, some algorithms to generate random variables (uniformly distributed variables, Gaussian variables, Gaussian pairs of correlated variables). We also treat the topic of characterizing Log-Normal variables [3, 8].

Joint PDFs, joint CDFs, conditional probabilities

The joint probability law of a set of two dependent variables (X, Y) is characterized by a joint CDF, a joint PDF, and Bayes conditional PDF, as outlined below:

$$F_{X,Y}(x, y) = \Pr(X \leq x, Y \leq y)$$

$$f_{X,Y}(x, y) = \frac{\partial^2 F_{X,Y}}{\partial x \partial y} \Leftrightarrow \begin{aligned} f_{X,Y}(x, y) dx dy &= d F_{X,Y}(x, y) \\ &= \Pr(x \leq X \leq x + dx, y \leq Y \leq y + dy) \end{aligned}$$

$$f_{Y|X}(y|x) = \frac{f_{X,Y}(x, y)}{f_X(x)} \tag{3.13}$$

Multivariate Gaussian distribution, correlations, and covariance matrix

Bivariate Gaussian distribution (two cross-correlated random variables)

If (X, Y) are jointly Gaussian, and cross-correlated, then we have the following bivariate Gaussian PDF:

$$f_{X,Y}(x,y) = \frac{1}{2\pi\sigma_X\sigma_Y}\exp\left\{-\frac{1}{2}\left(\frac{1}{1-\rho^2}\right)\left[\left(\frac{x-m_X}{\sigma_X}\right)^2 - 2\rho\left(\frac{x-m_X}{\sigma_X}\right)\left(\frac{y-m_Y}{\sigma_Y}\right) + \left(\frac{y-m_Y}{\sigma_Y}\right)^2\right]\right\}$$
(3.14)

Note that the covariance between the two variables is $\mathrm{Cov}(X,Y) = \rho\,\sigma_X\sigma_Y$, where ρ is the correlation coefficient between X and Y. This covariance appears more explicitly as an $N \times N$ covariance matrix in the more general multivariate case for N Gaussian random variables (shown further below). *Equation* 3.15 below summarizes covariance-correlation properties for 2 random variables, regardless of their joint probability law (Gaussian or not):

Covariance : $\mathrm{Cov}(X,Y) = \langle(X - m_X)(Y - m_Y)\rangle$ Correlation coefficient: $\rho = \rho_{XY} = \frac{\mathrm{Cov}(X,Y)}{\sigma_X\sigma_Y}$

Symmetric covariance matrix of size 2×2 : Symmetric covariance matrix of size 2×2 :

$$\underline{\underline{C}}_{XY} = \begin{bmatrix} \sigma_X^2 & \mathrm{Cov}(X,Y) \\ \mathrm{Cov}(X,Y) & \sigma_Y^2 \end{bmatrix} \qquad \underline{\underline{R}}_{XY} = \begin{bmatrix} 1 & \rho \\ \rho & 1 \end{bmatrix}$$
(3.15)

Multivariate Gaussian distribution (N cross-correlated Gaussian random variables)

The bivariate Gaussian distribution, defined earlier for 2 dependent Gaussian random variables (X, Y), can be generalized for $N > 2$ dependent (cross-correlated) random variables, named here $\{X_1, X_2, \ldots, X_N\}$, and represented as the random vector $\vec{X} = [X_1, X_2, \ldots, X_N]$. The joint multivariate Gaussian PDF of $\vec{X}$ is:

$$f_{\vec{X}}(\vec{x}) = \frac{1}{\sqrt{(2\pi)^N \det(C_{XX})}}\exp\left\{-\frac{1}{2}(\vec{x} - \vec{m}_X)^T C_{XX}^{-1}(\vec{x} - \vec{m}_X)\right\}$$
(3.16)

…where C_{XX} is the $N \times N$ symmetric covariance matrix of the N variables, containing the covariances $C_{ij} = \mathrm{Cov}(X_i, X_j)$ in row i and column $j \neq i$, and the variances $\mathrm{Var}(X_i)$ in the diagonal positions (i, i). Indeed, note that $C_{11} = \mathrm{Var}(X_1) = \sigma_{X1}^2$, $C_{22} = \mathrm{Var}(X_2) = \sigma_{X2}^2$, etc. The joint PDF given above in *Eq. 3.16* entirely describes the probability law of the multivariate Gaussian vector. It is remarkable that its joint probability law depends only on the N means m_{Xi}, the N variances σ_{Xi}^2, and the $N(N-1)/2$ symmetric cross-covariances $\mathrm{Cov}(X_i, X_j)$, and not on any higher order moments.

Uni- and multi-variate Log-normal distribution (and covariance matrix)

Recall that the support of a Gaussian variable is $]-\infty, +\infty[$: it can take negative values. Therefore, a set of positive parameters $\{X_1, \ldots, X_K\}$ with support $]0, +\infty[$ cannot be Gaussian, but they *could be* Log-Normal. It will be seen how the *cross-correlations* of Log-normals variables X_i's can be honored in the generation process, using moment relations between X_i and $Y_i = \mathrm{Ln}\, X_i$.

Univariate Log-normal law. Let us start by specifying the PDF for a single Log-normal variable, $X > 0$. First, recall that X (Log-normal) can be analyzed as the exponential of a Gaussian variable Y: $X = \exp(Y)$. One should be careful in clearly distinguishing the moments of the Log-normal variable X : $\mathrm{Log}\mathcal{N}(m_X, \sigma_X^2)$ and those of the Gaussian variable Y : $\mathcal{N}(m_Y, \sigma_Y^2)$. They are related as shown for instance in *Eq. 3.21* further below.

Let us first express the PDF of X in terms of moments of X (the positive variable):

$$f_X(x) = \frac{1}{x\sqrt{2\pi\,\mathrm{Ln}(1 + \sigma_X^2/m_X^2)}} \exp\left\{ -\frac{1}{2} \frac{\left[\mathrm{Ln}(x) - \mathrm{Ln}\left(m_X/\sqrt{1 + \sigma_X^2/m_X^2}\right)\right]^2}{\mathrm{Ln}(1 + \sigma_X^2/m_X^2)} \right\}$$

(3.17)

The PDF of X can alternatively be expressed using moments of the Gaussian $Y = \mathrm{Ln}(X)$, as follows.

The Log-normal PDF of $X > 0$, in terms of moments of the Gaussian $Y = \mathrm{Ln}(X)$:

$$f_X(x) = \frac{1}{x\sqrt{2\pi\,\sigma_Y^2}} \exp\left\{ -\frac{1}{2} \frac{[\mathrm{Ln}(x) - m_Y]^2}{\sigma_Y^2} \right\}$$

(3.18)

The two expressions above are equivalent. For instance, the coefficient of variation of X, $C_X \equiv \sigma_X/m_X$, verifies $C_X = \sqrt{\exp(\sigma_Y^2) - 1}$, which yields $\sigma_Y^2 = \mathrm{Ln}(1 + \sigma_X^2/m_X^2)$. Thus, the term $\mathrm{Ln}(1 + \sigma_X^2/m_X^2)$ in *Eq. 3.17* is identical to the term σ_Y^2 in *Eq. 3.18*.

Finally, the Log-normal CDF can be deduced from the Gaussian CDF using the *monotonic* transformation $X = \exp(Y)$. We have indeed:

$$F_X(x) = \mathrm{Proba}\{X \le x\} = \mathrm{Proba}\{\exp(Y) \le x\} = \mathrm{Proba}\{Y \le \mathrm{Ln}(x)\}$$
$$\Rightarrow \quad F_X(x) = F_Y(\mathrm{Ln}(x))$$

...where F_Y is the CDF of the Gaussian Y, expressed in terms of the *erf* function. Thus, we have finally obtained an explicit description of the Log-normal CDF of X (*here, in terms of moments of* $Y = \mathrm{Ln}X$):

$$\text{Log-normal CDF}: \quad F_X(x) = \frac{1}{2}\left\{ 1 + \mathrm{erf}\left(\frac{\mathrm{Ln}(x) - m_Y}{\sigma_Y\sqrt{2}} \right) \right\}$$

(3.19)

The reader may obtain another equivalent formulation of $F_X(x)$ in terms of moments of X, by using the relations between moments of X and moments of $Y = \mathrm{Ln}\,X$ (see below, *Eq. 3.21*).

Multivariate log-normal law

We now briefly characterize the joint PDF of a multivariate *Log-normal* vector of *positive* random variables, or random parameters. This is useful for Monte Carlo uncertainty analyses, where one needs to generate replicates of several cross-correlated *positive* parameters. Indeed, *positive* parameters cannot be Gaussian, but they can be assumed Log-Normal. It is therefore necessary to clarify how cross-correlations of Log-Normal parameters can be honored in the Monte Carlo simulation process. For instance, consider Log-normal input parameters X_1: *permeability,* X_2: *dispersivity (transverse),* X_3: *dispersivity (longitudinal),* X_4: *initial concentration* $(X_1 > 0; X_2 > 0; X_3 > 0; X_4 > 0)$. At least the first three can be cross-correlated. Their joint PDF can be expressed, as before, either in terms of moments of X_i's, or in terms of moments of the Gaussian Y_i's $\{Y_i = \mathrm{Ln}(X_i) \in \mathbb{R}\}$. The latter formulation is easier.

Let us consider for instance the bivariate case with 2 correlated Log-normals $\{X_1; X_2\}$:

Bi-variate Log-normal PDF of $\{X_1, X_2\}, \ldots$ in terms of

moments of the Gaussians $Y_j = \mathrm{Ln}(X_j)\{j = 1, 2\}$:

$$f_{X_1,X_2}(x_1, x_2)$$
$$= \frac{1}{2\pi\,\sigma_{Y1}\sigma_{Y2}}\frac{1}{x_1 x_2}\exp\left\{-\frac{1}{2}\left[\overrightarrow{\mathrm{Ln}(x)} - \overrightarrow{m_Y}\right]^T \mathbf{C}_{YY}^{-1}\left[\overrightarrow{\mathrm{Ln}(x)} - \overrightarrow{m_Y}\right]\right\} \qquad (3.20)$$

… with the vector notation : $\overrightarrow{\mathrm{Ln}(x)} = \begin{bmatrix}\mathrm{Ln}(x_1) \\ \mathrm{Ln}(x_2)\end{bmatrix}$ and $\overrightarrow{m_Y} = \begin{bmatrix}m_{Y1} \\ m_{Y2}\end{bmatrix}$.

The first term in the denominator can also be expressed as: $2\pi\ \sigma_{Y_1}\sigma_{Y_2} = \sqrt{(2\pi)^2 \det(\mathbf{C}_{YY})}$.

The moments of the Log-normals $\{X_1; X_2\}$ can be related as follows to those of the Gaussians $\{Y_1; Y_2\}$:

Means : $m_{X1} = \exp\left\{m_{Y1} + \frac{1}{2}C_{Y1Y1}\right\} = \exp\left\{m_{Y1} + \frac{1}{2}\sigma_{Y1}^2\right\}$;

$m_{X2} = $ similarly …

Variances−Covariances $(i = 1, 2; j = 1, 2)$:

$$\mathrm{Cov}(X_i, X_j) = \exp\left\{m_{Yi} + m_{Yj} + \frac{1}{2}\left(C_{YiYi} + C_{YjYj}\right)\right\}\left[\exp\left(C_{YiYj}\right) - 1\right]$$
$$\Rightarrow\ \mathrm{Cov}(X_i, X_j) = m_{Xi}m_{Xj}\left[\exp\left(C_{YiYj}\right) - 1\right]$$
$$\Rightarrow\ Var(X_i) = \sigma_{Xi}^2 = m_{Xi}^2\left[\exp\left(\sigma_{Yi}^2\right) - 1\right] \qquad (3.21)$$

For verification, the moment relation for the univariate case is known to be $\sigma_X^2 = m_X^2\left[\exp\left\{\sigma_Y^2\right\} - 1\right]$ with $Y = \mathrm{Ln}\ X$. This confirms the more general bivariate relation in *Eq. 3.21* above. In fact, *Eq. 3.21* provides the moment relations between Gaussian and Log-Normal for both the univariate case (take $i = j = 1$) and the bivariate case (take $i = 1, 2; j = 1, 2$).

In practice, with Monte Carlo simulations, where multiple replicates must be generated, one may avoid dealing with multivariate Log-normal $\{X_i's\}$, by generating instead multivariate Gaussian $\{Y_i's\}$. Taking the exponential of the $\{Y_i's\}$ replicates yields the Log-normal $\{X_i's\}$ replicates with $X_i = \exp(Y_i)$. The disadvantage, perhaps, is that moments of the transformed $Y_i = \text{Ln}(X_i)$ must first be inferred from moments of the natural positive variables X_i (*this is cumbersome but feasible using previous relations*).

Finally, it remains to be seen, technically, what *algorithms* can be used to generate multiple replicates of a cross-correlated multivariate vector for Monte Carlo simulations. For simplicity, we indicate below *generation algorithms* for two cross-correlated Gaussian parameters (Y_1, Y_2).

Algorithm to generate bivariate replicates of 2 cross-correlated parameters

Let $\{Y_1; Y_2\}$ be two Gaussian parameters with means, variances, and cross-covariance defined by:

$$Y_1 = \mathcal{N}(m_{Y1}, \sigma_{Y1}^2); \ Y_2 = \mathcal{N}(m_{Y2}, \sigma_{Y2}^2); \ \text{Cov}(Y_1, Y_2) = \rho \, \sigma_{Y1}\sigma_{Y2}$$

...where ρ is the *correlation coefficient* $(-1 \leq \rho \leq +1)$. The algorithm to generate a set of M replicates $\{m = 1, 2, \ldots, M\}$ of these bivariate Gaussian parameters is as follows.

Two cross-correlated Gaussians

First, let us assume that we have at our disposal an algorithm to generate, independently, M replicates of a univariate Gaussian variable $G_1 = \mathcal{N}(0, 1)$, and independently, M other replicates of a univariate Gaussian variable $G_2 = \mathcal{N}(0, 1)$, where G_1 and G_2 are independent, and both are normalized with zero mean and unit variance. Then we obtain M replicates of the pair of cross-correlated Gaussian variables $\{Y_1, Y_2\}$ using:

$$m = 1, 2, \ldots, M : \begin{cases} Y_1^{(m)} = m_{Y1} + \sigma_{Y1} G_1^{(m)} \\ Y_2^{(m)} = m_{Y2} + \sigma_{Y2}\left\{\rho \, G_1^{(m)} + \sqrt{1 - \rho^2} \, G_2^{(m)}\right\} \end{cases} \quad (3.22)$$

Equation 3.22 requires generating normal variates such as $G = \mathcal{N}(0, 1)$. For this purpose, one can either use available software functions, like RANDN in MATLAB, or else, one can apply the *Box-Muller* method. Briefly, let $G_1 = R \cos \theta$ and $G_2 = R \sin \theta$ where $\theta \in [0, 2\pi]$ is uniformly distributed in $[0, 2\pi]$, and $R > 0$ has a Rayleigh distribution with PDF $f_R(r) = r \times \exp(-r^2/2)$ and CDF $F_R(r) = 1 - \exp(-r^2/2)$. Then it can be shown that $\{G_1, G_2\}$ are two $\mathcal{N}(0, 1)$ independent normalized Gaussian variables. The following *Box-Muller* algorithm generates 2M independent replicates of $G = \mathcal{N}(0, 1)$:

$$m = 1, 2, \ldots, M : \quad \begin{cases} G_1^{(m)} = R^{(m)} \cos \theta^{(m)} \\ G_2^{(m)} = R^{(m)} \sin \theta^{(m)} \end{cases} \tag{3.23}$$

…to be inserted into *Eq. 3.22* to obtain the 2 cross-correlated Gaussian replicates $\left\{ Y_1^{(m)}, Y_2^{(m)} \right\}$.

This algorithm (*Eq. 3.23*) also requires (i) generating replicates of the uniform $\theta : U[0, 2\pi]$, and (ii) generating replicates of the Rayleigh variable R.

Rayleigh replicates for the Box-Muller generator

The Rayleigh variable R can be generated by the *Inverse CDF method*. Briefly, replicates $R^{(m)}$ $\{m = 1, 2, \ldots, M\}$ are obtained from the inverse CDF $F_R^{-1}(r)$ using a sequence of uniform $U[0, 1]$ random numbers $U^{(m)}$ as follows (e.g., [5: Sect. 5.15]: "The inverse problem"):

$$R^{(m)} = F_R^{-1}\left(U^{(m)}\right) \text{ or equivalently } F_R\left(R^{(m)}\right) = U^{(m)}$$

With $F_R(r) = 1 - \exp(-r^2/2)$, this yields the Rayleigh generation algorithm:

$$R^{(m)} = \sqrt{-2Ln\left(1 - U^{(m)}\right)} \quad \text{where } U^{(m)} \in [0, 1]$$

From statistical theory, the above sequence provides an unbiased sample of replicates $R^{(m)}$ having the required Rayleigh distribution.

Generation of uniform replicates U with multiplicative congruential random number generator

Finally, the uniform replicates $U^{(m)}$ could be generated from intrinsic functions like RAND available in Fortran, Matlab, etc. However, for a completely autonomous algorithm, independent of any software function, we indicate one possible method for generating uniform random numbers based on a congruential generator of random integers, which is then converted to uniform real numbers in [0, 1]. The algorithm is as follows:

(i) A pseudo-random integer $N_{RAND}(k)$ is generated at the kth iteration:

$$N_{RAND}(k) = (L \times N_{RAND}(k - 1) + c)(\text{modulo } M)$$

…where:

"c"	is a constant number (integer)
"M"	is called the modulus (integer),
"L"	is called the multiplier (integer),
"$N_{RAND}(0)$"	is the initial integer or "seed", also denoted N_0.

(b) A real uniform random variable $U \in [0.0, 1.0]$ is then obtained at each iteration as follows:

$$U(k) = N_{\text{RAND}}(k)/M$$

…where U(k) is the real-valued floating point result of division by two integers. Thus, in Fortran, the correct result is obtained by the instruction:

$$U(k) = \text{Float}(N_{\text{RAND}}(k))/\text{Float}(M)$$

Some generators can be computed in 32-bit arithmetic, like the one proposed in [6] but they have a short cycle (Cycle : $2^{18} = 262144$). We propose here another "equidistributed" generator studied by [2], and used for random field generation in [1], which needs 64-bit arithmetic (INTEGERS**8) and whose cycle is longer, more than two billion replicates (Cycle $= M - 1 = 2^{31} - 2 = 2.147483646 \times 10^9$):

Modulus : $M = 2^{31} - 1 = 2\,147\,483\,647$; Multiplier : $L = 950\,706\,376$;

Constant : $c = 0$; Seed : any integer from $1\ to\ M - 1$, for instance $N_0 = 1$

Cycle Length $= 2^{31} - 2 = 2\,147\,483\,646 \approx 2 \times 10^9$

Other multipliers L were judged satisfactory too; here is the list furnished by [2]:

$L = 950706376;\ 742938285;\ 1226874159;\ 62089911;\ and\ 1343714438$

…with a preference for the first multiplier $L = 950706376$.

Example: equiprobable generation of 2 Gaussian correlated parameters

The idea of *equiprobable sampling* of a set of two cross-correlated random parameters is illustrated in Fig. 3.3. A set of 1000 replicates of two correlated Gaussian variables was generated. The resulting cloud of points $\left(X^{(m)}, Y^{(m)}\right)$ is shown for $m = 1, 2, \dots, 250$ points. The red lines are the two regression lines $Y|X$ and $X|Y$, with the smallest absolute slope for $Y|X$, and the largest one for $X|Y$. From Bayesian estimation theory, $Y|X$ is the Best Linear Unbiased Estimator (BLUE) of Y conditioned by X, and $X|Y$ is the BLUE estimator of X conditioned by Y. This plot shows that the regression models $Y = a_{Y|X}X + b_{Y|X}$ and $X = a_{X|Y}Y + b_{X|Y}$ are not equivalent, because conditioning Y by X is not the same as conditioning X by Y. Iso-probability contours having the shape of ellipses are also shown in the plot of Fig. 3.3.

3.1.3 Estimating a Covariance or Correlation from a Data Sample

The previously defined covariances $C_{ij} = \text{Cov}\left(X_i, X_j\right)$ described the statistical relation between N random variables (N uncertain input parameters) taken two

Fig. 3.3 A sample of 1000 points (X, Y) was drawn from a negatively cross-correlated bivariate Gaussian distribution (only 250 points shown here). The red lines are the two regression lines $Y|X$ and $X|Y$. The orthogonal blue lines are the two principal diameters of iso-probability ellipses, computed from the joint probability law using empirical moments. For $N = 1000$, these were close to the theoretical moments: $\{m_X = 0, \sigma_X = 1\}$, $\{m_Y = 0, \sigma_Y = 2\}$, $\rho_{XY} = -1/2$

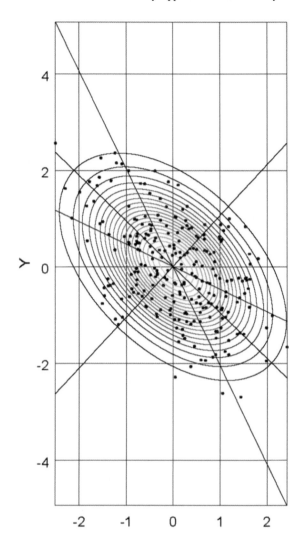

by two. These covariances characterize the dependence between the N variables regardless of their probability law (multivariate Gaussian or not). They can be used in fact to describe not only the dependence between several uncertain input parameters (e.g., based on field data), but also between several outputs of the model such as the pollutant concentrations obtained at different target points in space.

Covariance and correlation coefficient

Let us look briefly at the problem of estimating a covariance $\mathrm{Cov}(X, Y)$ from a finite data sample of the two variables. The following «statistic» is a consistent and unbiased estimator of the covariance $\mathrm{Cov}(X, Y)$ between two random variables X

and Y (two examples in our context: porosity Φ (m^3/m^3) and permeability k (m^2); transverse and longitudinal dispersivities α_T and α_L (m)):

$$\widehat{C}_{XY} = \frac{1}{N-1} \sum_{i=1}^{i=N} \left(X^{(i)} - \widehat{m}_X \right)\left(Y^{(i)} - \widehat{m}_Y \right) \tag{3.24}$$

The following «statistic» can then be proposed as a reasonable estimator (consistent *but* generally biased) of the cross-correlation coefficient ρ_{XY} between X and Y:

$$\widehat{\rho}_{XY} = \frac{\frac{1}{N-1}\sum_{i=1}^{i=N}\left(X^{(i)}-\widehat{m}_X\right)\left(Y^{(i)}-\widehat{m}_Y\right)}{\sqrt{\left(\frac{1}{N-1}\sum_{i=1}^{i=N}\left(X^{(i)}-\widehat{m}_X\right)^2\right)\left(\frac{1}{N-1}\sum_{i=1}^{i=N}\left(Y^{(i)}-\widehat{m}_Y\right)^2\right)}} \tag{3.25}$$

Linear regression and correlation coefficient

The correlation coefficient also plays a crucial role in linear regression models $Y|X$, where Y is estimated (explained) from given X values (explanatory variable). The linear regression model yields:

$$Y(X) = aX + b + \varepsilon,$$

where the slope a $= \rho_{XY}\, \sigma_Y/\sigma_X$ minimizes error variance $\mathrm{Var}(\varepsilon)$, and $b = m_Y - a\, m_X$ is obtained from the unbiasedness condition $E(\varepsilon) = 0$. In practice, parameters a and b are estimated by replacing all quantities by their estimates ■ (*estimators were defined earlier*). An important result in practice is the minimal error variance, which can be used to obtain a scaled measure of root-mean-square error:

$$\sigma_\varepsilon^2 = \sigma_Y^2\left(1 - \rho_{XY}^2\right) \rightarrow e_{\mathrm{RMS}} = \sigma_\varepsilon/\sigma_Y = \sqrt{1 - \rho_{XY}^2} \text{ (between 0 and 100\%).}$$

Examples : $\quad \rho_{XY} = 0.70 \rightarrow e_{\mathrm{RMS}} \approx 71\%$; $\quad \rho_{XY} = 0.95 \rightarrow e_{\mathrm{RMS}} \approx 31\%$

This shows (*perhaps surprisingly to some readers*) that even with a "good" correlation coefficient 0.95, the root-mean square error is still 31%: thus, in this case, only 69% of the variability of Y is "explained" by the linear regression model $Y(X)$.

References

1. R. Ababou, L.W. Gelhar, D. McLaughlin, Three-dimensional flow in random porous media, 2 vols. Technical report No. 318, Ralph Parsons Laboratory for Water Resources & Hydrodynamics, Department of Civil Engineering, Massachusetts Institute of Technology (MIT), Cambridge, Massachusetts, USA, March 1988 (1988), 833pp., http://hdl.handle.net/1721.1/14675
2. G.S. Fishman, L.R. Moore, An exhaustive analysis of multiplicative congruential random number generators with modulus (2**31)-1. SIAM J. Sci. Stat. Comput. **7**, 24–45 (1986)
3. L.J. Halliwell, The Lognormal Random Multivariate. Casualty Actuarial Soc. E-Forum, Spring **5** (2015)
4. L. Isserlis, On a formula for the product-moment coefficient in any number of variables. Biometrika **12**(1–2), 134–139 (1918)
5. A. Papoulis, S.U. (Unnikrishna) Pillai, in *Probability, Random Variables, and Stochastic Processes* (16 Chaps.), 4th edn. (Mc-Graw Hillition, 2002), 852pp.
6. A.F.B. Tompson, R. Ababou, L.W. Gelhar, Implementation of the three-dimensional turning bands random field generator. Water Resour. Res. **25**(10), 2227–2243 (1989)
7. E. Vanmarcke, *Random Fields: Analysis and Synthesis* (Massachusetts Institute of Technology Press, Cambridge, Massachusetts, 1983), p.382
8. Wikipedia, Wikipedia article "Log-normal distribution" (English) (2022). https://en.wikipedia.org/wiki/Log-normal_distribution (web page, last edited 02 Sep. 2022, downloaded 11 Oct. 2022)
9. L.A. Zadeh, Outline of a new approach to the analysis of complex systems and decision processes. IEEE Trans. Syst. Man Cybern. SMC-**3**, 28–44 (1973)
10. L.A. Zadeh, Fuzzy sets. Inf. Control **8**(3), 338–353 (1965)

Chapter 4
Fuzzy Set Characterization
of Uncertainty (Fuzzy Variables)

Fuzzy set theory (fuzzy variables & fuzzy logic) emerged in 1965 with L. A. Zadeh's article on "Fuzzy Sets," followed by another article [1]. There have been precursors before these initial works, and further extensions of fuzzy set theory afterwards and up to the recent years. In particular, the initial concepts of fuzzy logic were extended (or refined) based on the theory of "possibilities" [2]. More recently, the theory of possibilities was used to combine probabilistic and fuzzy logic representations of uncertain variables, and to propagate fuzzy uncertainty by various methods such as the technique of Independent Random Sets (e.g., [3]), to be presented further below. Let us first present an overview of various concepts from fuzzy set theory, and probabilistic extensions of it like the theory of possibilities combining fuzziness and randomness. We will then review a few applications, algorithms, and softwares based on fuzzy variables.

4.1 Fuzzy Sets, Fuzzy Numbers, Fuzzy Logic and Arithmetic

4.1.1 Fuzzy Sets, Fuzzy Numbers, Membership Functions, and α-Cuts

Briefly, a variable $X \in A \subset \mathbb{R}$ is fuzzy, or belongs to the fuzzy set A, if the value taken by X in this set is only defined to some degree $\mu(X)$ comprised between zero (0) and one (100%). That is, for instance, we may have $X = 3.14$ with degree $\mu(3.14) = 0.75 = 75\%$. We have in this case $X = 3.14$ with a truth degree of 75%, and $X \neq 3.14$ with a truth degree of 25%.

Considering now the whole set $A \subset \mathbb{R}$ on which X is defined, the degree function $\mu(X)$ can be specified for all values $X \in A$. This function $\mu(X)$ is also called

© The Author(s), under exclusive license to Springer Nature Singapore Pte Ltd. 2023
R. Ababou et al., *Uncertainty Analyses in Environmental Sciences and Hydrogeology*,
SpringerBriefs in Applied Sciences and Technology,
https://doi.org/10.1007/978-981-99-6241-9_4

the membership function of the fuzzy variable X, because $\mu(X)$ is the degree of membership of $X \in A$. The membership function $\mu(X)$ can also be described in terms of its α-cuts, set of intervals such that $\mu(X) = \alpha$ with $\alpha \in [0, 1]$. For a concave membership functions, there is a unique α-cut interval $A_{\alpha} = [X_{\text{LEFT}}(\alpha), X_{\text{RIGHT}}(\alpha)]$ for each value of α; examples: triangular, trapezoidal, Gaussian, or also, the trivial box-shaped membership function. The α-cut interval A_0 corresponding to $\alpha = 0$ is the set of all values X for which it is possible that X satisfies this value (or this property). Indeed, for values outside this α-cut, we have $\mu(x) = 0$, which means that these values cannot satisfy the desired property (the desired value of X).

Here are three examples of membership functions, taking for example $X = \Phi$ (porosity) and $A = [0, 1]$. Note: the three shapes below (box, triangle, trapeze) could all be described as generalized trapezoidal functions, since the box and the triangle are just special cases of the trapeze. For theoretical indications on the practical interest of these membership functions, see [4].

- Box-shaped membership function: $\mu(\Phi) = 1$ on $X \in [\Phi_{\text{LOW}}, \Phi_{\text{UP}}] \subset [0, 1]$, and $\mu(\Phi) = 0$ outside. *Practical meaning*: we only know, or we think we know from expert judgment, that Φ cannot be less than Φ_{LOW} and cannot be greater than Φ_{UP}.
- Triangular-shaped membership function: $\mu(\Phi) = 1$ at $\Phi = \Phi_{REF}$, the apex or "core" of $\mu(\Phi)$; $\mu(\Phi)$ is zero for $\Phi \leq \Phi_{\text{LOW}}$ (at left) and for $\Phi \geq \Phi_{\text{UP}}$ at right. This membership function is formalized by the triplet $\{\Phi_{\text{LOW}}; \Phi_{\text{REF}}; \Phi_{\text{UP}}\}$. *Practical meaning*: porosity is a fuzzy number having an apex Φ_{REF} (reference value) and a support interval $[\Phi_{\text{LOW}}, \Phi_{\text{UP}}]$ (bounds).
- Trapezoïdal shaped membership function taking the value $\mu(\Phi) = 1$ within an inner interval $[\Phi_{\text{LowRef}}; \Phi_{\text{UpRef}}]$, and vanishing at the left of Φ_{Low} and at the right of Φ_{Up}; this trapezoïdal function can be represented as the quadruplet $\{\Phi_{\text{Low}}; \Phi_{\text{LowRef}}; \Phi_{\text{UpRef}}; \Phi_{\text{Up}}\}$.

In particular, the triangular function with $\mu(X_{\text{REF}}) = 1$ defines a so-called "fuzzy number", X_{REF}, which is "crisp" (the opposite of fuzzy), while neighboring values around $X = X_{\text{REF}}$ are fuzzy ($\mu < 1$). More generally, membership functions can have different shapes, and they could be strictly "fuzzy" (non-crisp) everywhere, that is, with $\mu(X) < 1, \forall X \in A$.

4.1.2 Fuzzy Logic and Fuzzy Arithmetic Rules in Models

One possible way to implement input/output model calculations with fuzzy input parameters in subsurface flow and transport simulations is to use fuzzy logic rules $(\cup, \cap, \Rightarrow, \Leftrightarrow, \sim)$, that is *(Or, And, Imply, Equivalent, Not)*, and also, fuzzy arithmetic rules $(+, -, \times, /)$.

For simplicity, consider for instance two fuzzy sets A and B with *symmetrical* triangular membership functions $\mu_A(X)$ and $\mu_B(X)$. Because of their symmetry, they can be defined by doublets $\{a_{\text{LOW}}; a_{\text{UP}}\}$ and $\{b_{\text{LOW}}; b_{\text{UP}}\}$, or better, by the intervals

$[a_{\text{LOW}}, a_{\text{UP}}]$ and $[b_{\text{LOW}}, b_{\text{UP}}]$. Arithmetic rules and logical rules are then as shown below, at least for fuzzy sets with *symmetric triangular membership functions*.

These rules were implemented for instance in the Fuzzy Logic groundwater flow code FLO2SIM of [5]: see Sect. 4.2.4 further below. The subtraction rule could serve for instance to calculate the fuzzy hydraulic head difference ΔH between spatial cells. More extended versions of fuzzy differential and integral calculus have been developed in the literature, but they were not used in FLO2SIM and they are beyond the scope of the present work.

Arithmetic rules of fuzzy calculus

Addition of fuzzy sets: $A \oplus B = [a_{\text{LOW}} + b_{\text{LOW}}, a_{\text{UP}} + b_{\text{UP}}]$.

Subtraction of fuzzy sets: $A \ominus B = [a_{\text{LOW}} - b_{\text{UP}}, a_{\text{UP}} - b_{\text{LOW}}]$.

Multiplication of fuzzy sets: $A \otimes B = \left[\text{Min}\{a_{\text{LOW}}b_{\text{LOW}}; a_{\text{LOW}}b_{\text{UP}}\}, \right.$

$\left. \text{Max}\{a_{\text{UP}}b_{\text{LOW}}; a_{\text{UP}}b_{\text{UP}}\} \right]$.

Division of fuzzy sets: $A \oslash B = \left[\text{Min}\{a_{\text{LOW}}/b_{\text{LOW}}; a_{\text{LOW}}/b_{\text{UP}}\}, \right.$

$\left. \text{Max}\{a_{\text{UP}}/b_{\text{LOW}}; a_{\text{UP}}/b_{\text{UP}}\} \right]$.

Inverse of a fuzzy set: $A^{-1} = [1/a_{UP}, 1/a_{LOW}]$.

These rules are not entirely obvious: note that $A \ominus A \neq [0, 0]$ and that $A \oslash A \neq [1, 1]$.

Logical rules of fuzzy calculus

Union of fuzzy sets: A or $B = A \cup B = \text{Max}\{A; B\}$.

Intersection of fuzzy sets: A and $B = A \cap B = \text{Min}\{A; B\}$.

The intersection of 2 non-overlapping triangular membership functions is empty (the resulting membership function is null everywhere): for instance, with symmetric triangular functions "μ", we have $\mu_{[0.5,3.5]} \cap \mu_{[7.0,9.5]} = \emptyset$.

Figure 4.1 illustrates the fuzzy logic "union" rule, which calculates the membership function $\mu_{A\cup B}(X)$ of the union of two fuzzy sets A and B having triangular memberhip functions $\mu_A(X)$ and $\mu_B(X)$. In this example, each membership function is triangular symmetric. The union rule can be extended more generally to any proper membership functions other than triangular.

The union $\cup$ ("or") of 2 triangular fuzzy sets can yield a non-triangular fuzzy set. Thus, in Fig. 4.1, it can be seen that the union of 2 overlapping triangles can yield an M-shaped membership function (not triangular). In fact, the fuzzy outputs of a model can have complicated membership functions even if the fuzzy inputs are all triangular: see example from the FLO2SIM model, shown further below in Fig. 4.7 in Sect. 4.2.4.

Remark. Even if the model is based on linear PDE's, it can be nonlinear in terms of input/output dependence, whence nonlinear fuzzy logic and arithmetic

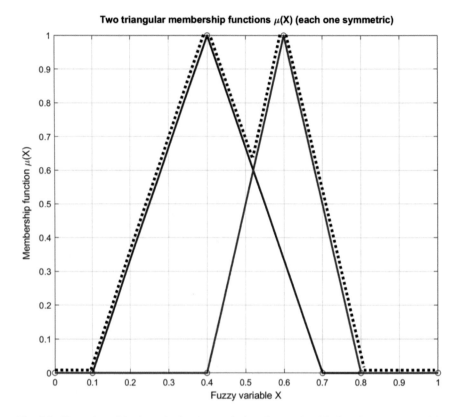

Fig. 4.1 Illustration of the fuzzy logic rule to calculate the membership function $\mu_{A \cup B}(X)$ of the union of 2 fuzzy sets A and B having triangular memberhip functions $\mu_A(X)$ and $\mu_B(X)$. The black dashed curve represents the function $\mu_{A \cup B}(X)$ of the union $A \cup B$

rules. *Example:* the FLO2SIM model discussed in Sect. 4.2.4 which implements nonlinear fuzzy rules to quantify fuzzy outputs from fuzzy inputs. The model is based on a linear PDE combining mass conservation and the linear law of Darcy $\vec{q} = -K(x, y)\overrightarrow{\text{grad}}(H)$ where the gradient is approximated by finite differences.

4.2 From Fuzzyness to "Possibility" (Fuzzy + Probabilistic)

This subsection is a brief description of a relatively recent extension of fuzzy set theory to include uncertainty on the membership function of the fuzzy variables. The earliest reference on this extension is [2] *(the same article bears two different dates in the literature),* we also follow the more recent presentations of [3, 6, 7]. The theory of "*possibilities*" was developed by [2], under the title *«Fuzzy sets as a basis for a theory of possibility».* That work extended and refined the previously developed theory of fuzzy variables (or fuzzy sets). It clarified formally the relation between fuzzyness and probability, which can be summed up as follows:

$$Necessity < Probability < Possibility.$$

4.2.1 Fuzzy Probability Bounds: Possibility (Upper CDF), Necessity (Lower CDF)

A probability distribution, that is, a Cumulated Distribution Function (CDF), is used to represent a quantity whose value depends on chance (random variability). Similarly, a distribution of "possibilities" can represent, for a given variable, an information that is incomplete or imprecise (i.e., a situation of partial knowledge, or partial ignorance). In fact, it appears that the notion of "possibility" can be interpreted as a "degree of possibility" for the given variable, which is analogous to the notion of "likelihood" in the classical fuzzy set theory. Note also that the likelihood function is another name for the membership function $\mu(X)$ of the fuzzy variable X, which is shown for instance further below in Fig. 4.2. Note also that the membership function characterizes the degree of "crispness" of X. Thus $\mu(X) = 0$ is associated to total "fuzziness," while $\mu(X) = 1$ is associated to total "crispness."

More recently, in their article on the *«Joint propagation and exploitation of probabilistic and possibilistic information in risk assessment models»,* [3] introduced the technique of *Independent Random Sets* (IRS), which combines probability and fuzzy logic via the theory of possibilities (among other similar approaches). Such techniques can be viewed as a means of representing incomplete or imperfect knowledge of the probabilistic distribution of one or several random variables. Furthermore, in some theoretical works, where likelihood functions are studied in a probabilistic & possibilistic setting, membership functions of fuzzy variables can be interpreted as ("coherent") conditional possibilities: see for instance [8], and references therein.

To sum up, the theory of "possibilities" (and "necessities") appears as a hybrid probabilistic extension of fuzzy logic. A practical presentation of it was provided in the report by [6], who considered for simplicity the "pH" as a fuzzy possibilistic variable "X" (e.g., in a geochemistry or water chemistry model). Their interpretation of the likelihood function of the fuzzy variable X, based on "possibility" theory, is expressed in terms of probability functions (CFDs). This type of interpretation is indicated in Fig. 4.2, taking the example of porosity: $X = \Phi$, with $\Phi \in [0, 1]$. We start with a trapezoïdal membership function for the fuzzy or imprecise variable X (porosity); this function is bounded by two CDFs, interpreted as follows within the theory of "possibilities":

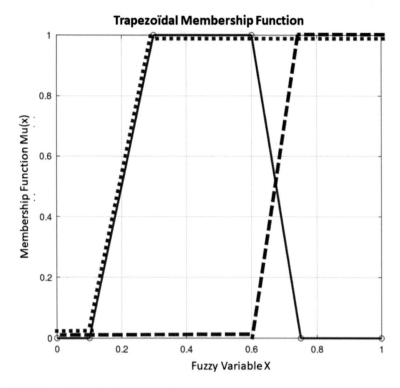

Fig. 4.2 Theory of "possibilities": trapezoïdal membership function $\mu(X)$ of a fuzzy (imprecise) variable, represented with two bounding Cumulated Distribution Functions (CDFs). The ordinate between 0 and 1 represents the degree of membership (blue curve) and the two CDFs (thick dashed curves). The *upper bound* CDF at left (short red dashes) is the "Possibility" function. The *lower bound* CDF at right (long magenta dashes) is the "Necessity" function. The abscissa represents the imprecise or fuzzy variable X, for example $X = \Phi$ (porosity)

> The *upper bound* CDF at left (red dashes) is interpreted as a measure of "Possibility" «$\Pi(X)$».
> The *lower bound* CDF at right (magenta dashes) is viewed as a measure of "Necessity" «$N(X)$».

Each of these two Cumulated Distribution Functions ($\Pi(X)$ and $N(X)$) expresses the probability that the porosity variable (Φ) be less or equal than the given value X. Indeed, classically, a CDF is a probability of non-exceedance. Thus, if we label $F_\Phi(X)$ the CDF of porosity Φ, we have classically for a probabilistic porosity: $F_\Phi(X) \equiv \mathrm{Proba}\{\Phi \leq X\}$. However, for a fuzzy porosity, as shown in Fig. 4.2, there is a fuzzy membership function $\mu(\Phi)$ characterizing the "crispness" of Φ and, in addition, there are two distinct CDFs characterizing bounds regarding the knowledge (or lack of knowledge) concerning Φ.

In summary, it can be seen from this example that the probabilistic CDF characterizing the porosity variable Φ is itself imprecise; the CDF of Φ is bound at left by the possibility measure «$\Pi(X)$» (*upper bound*), and at right, by the necessity measure «$N(X)$» (*lower bound*). Observe that:

- The "true" probability measure of Φ (its true CDF) remains unknown. We only know that the true CDF is located between these two bounds, «$\Pi(X)$» and «$N(X)$» shown in Fig. 4.2.
- The discrepancy between «$\Pi(X)$» and «$N(X)$» is a measure of our ignorance: the larger the discrepancy $N(X) - \Pi(X)$, the larger our ignorance concerning the porosity variable $X = \Phi$.

A final remark is in order concerning the CDF's denominations *"lower bound"* and *"upper bound"*, which is correct but may sound counter-intuitive. Consider the left CDF, named *Possibility* measure $\Pi(x)$. Recall that this CDF, $\Pi(x) = \text{Proba}(X \leq x)$, is the probability of non-exceedance of the variable X. Similarly, the right CDF, named *Necessity* measure $N(X)$, is the non exceedance probability $N(X) = \text{Proba}\{X \leq x\}$ with a different probability law than the previous one. Comparing the two functions in the porosity example of Fig. 4.2, it can be seen that we always have $N(x) \leq \Pi(x)$ for any fixed value x of the variable X.

Looking at Fig. 4.2, we have for instance:

$$x = 0 : N(x) = 0 \leq \Pi(x) = 0$$

$$x \approx 0.2 : N(x) = 0 \leq \Pi(x) \approx 0.5$$

$$x = 0.4 : N(x) = 0 \leq \Pi(x) = 1$$

$$x \approx 0.7 : N(x) \approx 0.7 \leq \Pi(x) = 1$$

$$x = 0.8 : N(x) = 1 \leq \Pi(x) = 1$$

More generally, for a possibilistic variable X, it can be shown [3] that its probability of non-exceedance of a given threshold value x can be bounded by the necessity function $N(x)$ as lower bound CDF, and the possibility function $\Pi(x)$ as upper bound CDF:

$$N(x) \leq \text{Proba}\{X \leq x\} \leq \Pi(x). \tag{4.1}$$

4.2.2 Intermediate CDF's via α-Cuts (Confidence Levels and Intervals)

The earlier discussion around Fig. 4.2 on possibility measure as an *upper bound* probability CDF, and necessity measure as a *lower bound* probability, deserves further remarks:

(1) The description around Fig. 4.2 of the upper and lower bound CDFs (possibility and necessity) can be further refined by introducing one or several intermediate probability curves (intermediate CDFs), each associated to a given confidence level α (more on this later).
(2) IRS theory (Independent random Sets) is a technique to calculate the propagation of uncertainty through an input/output model based on the theory of possibilities.
(3) In IRS theory, the 2 bounding CDFs, Possibility $\Pi(X)$ at left (*upper bound*), and Necessity $N(X)$ at right (*lower bound*), are also named respectively «Plausibility function» (at left) and «Belief Function (at right), although we will rather name them Possibility and Necessity here.

Accordingly, to complete the previous description, let us focus briefly on the concept of α-cuts.

Let A be the set of values of the variable X, for instance $A = [0, 1] \subset \mathbb{R}$, and let $\mu_A(X)$ be the triangular membership function of X defined on A. Let us take as an example the triangular membership function:

$$\mu_A(X) = \begin{vmatrix} (X - \ell)/(m - \ell) \text{ if } \ell \leq X \leq m \\ (r - X)/(r - m) \text{ if } m \leq X \leq r \end{vmatrix} \tag{4.2}$$

...where the triangle $[\ell, m, r]$ is defined by its left, middle, and right points. The following equation can then be used to define a set of α-cuts levels and intervals on the triangular membership function:

$$Cut_\alpha(A) = \{X | \mu_A(X) \geq \alpha\}; \ X^{(\alpha)} = [\ell + \alpha(m - \ell), r - \alpha(r - m)] \tag{4.3}$$

...for any confidence level $\alpha \epsilon [0, 1]$.

A similar definition of α-cuts levels and intervals can be formulated for trapezoidal membership functions, as shown graphically in Fig. 4.3. In total, $L + 1$ cuts can be defined with $\alpha_j\{j = 0, 1, \ldots, L - 1, L\}$, comprising the two extreme cuts plus $(L - 1)$ intermediate α-cuts $\alpha_j\{j = 1, \ldots, L - 1\}$.

Thus, Fig. 4.3 depicts a "possibilistic" representation of a fuzzy variable with a trapezoïdal membership function and three α-cuts (comprising just one intermediate cut). The graph can be interpreted as a distribution of possibilities for the imprecise parameter $X = C$, where "C" could be for instance a solute concentration at some boundary. The left ordinate axis in Fig. 4.3 represents a degree of possibility $\Pi(X)$ (which generalizes the usual membership function of classical fuzzy variables). The

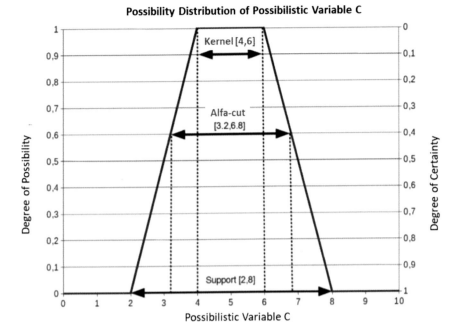

Fig. 4.3 "Possibilistic" representation of a fuzzy or "imprecise" variable $X = C$ with α-cuts. It illustrates a distribution of possibilities for C (e.g., solute concentration at a boundary). The left ordinate is the degree of possibility $\Pi(X)$, and the right ordinate is the degree of necessity $N(X)$. Three α-cuts are shown, with embedded confidence intervals (intermediate interval: [3.2, 6.8])

right ordinate represents the "complementary" degree of certainty, or degree of necessity $N(X)$. The three α-cuts define three embedded confidence intervals comprising the two left and right CDF's shown in the previous figure, plus one intermediate CDF not shown here.

Figure 4.3 illustrates the following expert opinion: *"The expert is sure that the uncertain parameter C (e.g. boundary concentration) is between 2 and 8, but values lying between 4 and 6 are the most likely."*

4.2.3 Random + Fuzzy Uncertainty: IRS Algorithm for Propagating Possibilistic CDFs

The goal of "possibilistic" uncertainty analysis is to propagate, through an input/output model, the collection of probability CDFs associated with α-cuts $\alpha_j \{j = 0, 1, 2, \ldots, L\}$. One way to implement this combined random/fuzzy uncertainty propagation is the IRS technique (*Independent Random Sets*), which we summarize below. For more details, see the paper on *"Joint propagation and exploitation of probabilistic and possibilistic information in risk assessment models"* by [3].

Note that another somewhat related variant of combined random + fuzzy propagation algorithm, the so-called *"Hybrid approach for addressing uncertainty in risk assessments,"* is described by [7], *and also by* [9].

In fact, the "hybrid propagation technique" and the "Independent Ransom Sets (IRS)" techniques are essentially two variants from a class of methods that seek to propagate random + fuzzy uncertainty through a model. The two variants (Hybrid and IRS) are presented in [9] in *their* sections [4-1] and [4-2]. Four variants are reviewed and compared in [10]: *the* IRS method, the "Hybrid" approach, plus two other methods: "Conservative Random Sets," and "Dependency Bounds Convolution." The "Hybrid" method processes random variability and fuzziness (imprecision) separately, combining Monte Carlo sampling with extended fuzzy set theory. The *R-Project* package HYRISK [11] implements this "Hybrid Approach."

The "Independent Random Sets" (IRS) method processes variability and fuzziness (imprecision) in the framework of Belief functions and Plausibility functions—or equivalently—Possibility functions $\Pi(X)$ (*upper bound CDF's*) and Necessity functions $N(X)$ (*lower bound CDF's*). In IRS, Monte Carlo sampling is applied both to the random probabilistic parameters (according to their probability distribution) and to the fuzzy possibilistic parameters (according to their membership or possibility function). The qualifier *"Independent"* stems from the fact that the method assumes *independence* between all sources of uncertainty, whether random ("aleatoric") or fuzzy/possibilistic ("epistemic"). Technically, the IRS algorithm selects replicates of the different random parameters independently from each other, and it also selects the α-cuts of the fuzzy variables independently of each other and independently of the random replicates. (The other variant, called "Hybrid method" or "Fuzzy Monte Carlo," is based on a similar procedure but the details of the sampling differ from IRS).

Let us now focus on the IRS method in more detail. We first define the following uncertain input variables or parameters of the model:

$\{X_1, X_2, \ldots, X_N, X_{N+1}, \ldots, X_{N+K}\}$: set of $N + K$ uncertain variables (model inputs).

We now separate them in order to distinguish between the random parameters and the fuzzy ones (it will also be useful to rename them, as indicated below):

$\{X_1, X_2, \ldots, X_N\} = \{P_1, P_2, \ldots, P_N\}$: sub-set of N random parameters "P_n" (purely probabilistic).
$\{X_{N+1}, \ldots, X_{N+K}\} = \{\Phi_1, \ldots, \Phi_K\}$: sub-set of K fuzzy parameters "Φ_k" (imprecise, possibilistic).

Therefore, with this notation, the set of uncertain input parameters is:

$$\{P_1, P_2, \ldots, P_N; \Phi_1, \ldots, \Phi_K\}$$

The N random parameters $\{P_1, P_2, \ldots, P_N\}$ are characterized by their N probability CDFs $\{F_{P1}, \ldots, F_{PN}\}$ or, if they are dependent, by their N-dimensional joint probability CDF $(F_{P1,\ldots,PN})$. On the other hand, the K imprecise or fuzzy parameters $\{\Phi_1, \ldots, \Phi_K\}$ are represented by their K possibility distributions $\{\Pi_1(\Phi_1), \ldots, \Pi_K(\Phi_K)\}$. These will be sampled discretely by α-cuts leading to fuzzy intervals labeled "$A_k^\alpha\{\ldots\}$". The chosen α-cuts are performed on triangular or trapezoïdal membership functions (these functions are also considered as degrees of possibility).

For any set of parameters and fuzzy variables $\{P_1, P_2, \ldots, P_N; \Phi_1, \ldots, \Phi_K\}$, the model outputs are a function of these inputs. For simplicity, consider a single output of interest, and call $\mathcal{M}\{\ldots\}$ the model functional that yields the output response "R" of interest; we express this input/output relation as:

$$R = \mathcal{M}\{P_1, P_2, \ldots, P_N; \Phi_1, \ldots, \Phi_K\} \tag{4.4}$$

In the IRS technique, the key point is that both types of uncertainties, random (probabilistic) and fuzzy (possibilistic), are propagated through the model. The technique is to first randomly sample the CDFs of the random parameters "P_n" repeatedly (Monte Carlo simulation), and to select (also repeatedly) the α-cuts (intervals) of the fuzzy possibilistic variables Φ_k from their possibility measure $\Pi_k(\Phi_k)$. The final step consists in finding *optimally* the upper and lower bounds of the model output, based on the response of the model for the entire set of sampled uncertain variables $\{P_1, P_2, \ldots, P_N; \Phi_1, \ldots, \Phi_K\}$ (replicated M times with M Monte Carlo samples).

The IRS propagation algorithm is detailed in Fig. 4.4, and illustrated graphically in Fig. 4.5.

Note that there are two model responses, R_{LOW} and R_{UP}. Because of the random Monte Carlo sampling of both the probabilistic parameters $\{P_n\}$ and the α-cut intervals $\{A_k\}$, both R_{LOW} and R_{UP} are random. They are characterized by lower and upper probability distributions. The final results, after stopping the Monte Carlo loop in the algorithm of Fig. 4.4, are the constructed lower and upper CDFs, $F_R^{\text{LOW}}(r)$ and $F_R^{\text{UP}}(r)$, of the model response. These two CDF's are different due to the fuzzy parameters $\{\Phi_k\}$. If these were absent, with only the random parameters $\{P_n\}$ remaining, the result would be a single CDF $F_R(r)$ of the model response R.

Application of IRS uncertainty analysis. We will present later in Sect. 5.4.2 an example application of the IRS technique using a simplified 1D concentration migration model, based on the theory of possibilities combining fuzzy and probabilistic variables as reviewed here.

Remark on possibilistic Monte Carlo simulations versus Fuzzy Logic. In the IRS technique, Monte Carlo simulations are implemented. That is, random sampling is performed. Consequently, low probability values are missed if the sample is small. This is different from classical fuzzy logic, which does not necessarily require Monte Carlo sampling for uncertainty propagation, as observed by [12]. Applications of *fuzzy logic* are reviewed later in Sect. 4.2.4.

The set of uncertain parameters (random & fuzzy) is denoted $\{X_1, X_2, \ldots, X_N, X_{N+1}, \ldots, X_{N+K}\}$ or equivalently $\{P_1, P_2, \ldots, P_N ; \Phi_1, \ldots, \Phi_K\}$.

- The N probabilistic random parameters are assumed independent, and they are characterized by their univariate probability distributions, the CDF's $\{F_{P1}, \ldots, F_{PN}\}$.

- The K fuzzy or "possibilistic" variables $\{\Phi_1, \ldots, \Phi_K\}$ are characterized by their K possibility distributions $\{\Pi_1(\Phi_1), \ldots, \Pi_K(\Phi_K)\}$.

(1) Monte Carlo steps $m = 1, \ldots, M$ (loop):

At each new step m, generate $N + K$ uniformly distributed random numbers in $[0,1]$: $\{u_1, \ldots, u_N\}^{(m)}$ for the N random parameters, and $\{\alpha_1, \ldots, \alpha_K\}^{(m)}$ for the K fuzzy variables.

(2) Monte Carlo generation of the probabilistic input parameters

Use $\{u_1, \ldots, u_N\}^{(m)}$ for equiprobable Monte Carlo generation (sampling) of the N random parameters P_n independently of each other, according to their probability distributions. This is performed by the inverse CDF method explained earlier:

$$P_n^{(m)} = F_{Pn}^{-1}\left(u_n^{(m)}\right) \text{ for the N parameters } P_n \ (n = 1, \ldots, N)$$

(3) Monte Carlo generation of α-cut intervals for the fuzzy input parameters

At each Monte Carlo step m, use the random levels $\{\alpha_1, \ldots, \alpha_K\}^{(m)}$ to sample K intervals $(\alpha - cuts)$ for the K fuzzy variables Φ_k. That is, applying to each fuzzy variable Φ_k the cut $\Pi_k(\Phi_k) = \alpha_k^{(m)}$ generates the $(\alpha - cut)$ intervals $A_k^\alpha \equiv A\left\{\Pi_k(\Phi_k) = \alpha_k^{(m)}\right\}$.

Note(1): the $\alpha - cuts$ are performed independently of each other (because the α_k's are independent), and they are performed independently from the random parameters (because the α_k's are also independent from the u_n's).

Note(2): in some applications, e.g. failure & reliability analyses, the $\alpha - cuts$ are also used to define the regions above the α-levels, that is the α-regions: $A_k^\alpha \equiv A\left\{\Pi_k(\Phi_k) \geq \alpha_k^{(m)}\right\}$.

(4) Calculate the smallest (Inf) and largest (Sup) values of the model response

The model response R is predefined (as explained earlier). At each step m of the Monte Carlo loop, the model response is of the form: $R^m = \mathcal{M}\left\{P_1^{(m)}, \ldots, P_N^{(m)}; \Phi_1, \ldots, \Phi_K\right\}$. Since the fuzzy variables are represented by their $\alpha - cut$ intervals A_k^α defined above, the model response to be analyzed takes the form: $R^{(m)} = \mathcal{M}\left\{P_1^{(m)}, \ldots, P_N^{(m)}; A_1^\alpha, \ldots, A_K^\alpha\right\}$. To calculate the smallest (Inf) and largest (Sup) values of the model response $R^{(m)}$, interval analyzis should be applied, using all the values located within the $(\alpha - cut)$ intervals $A_k^\alpha \equiv A\left\{\Pi_k(\Phi_k) = \alpha_k^{(m)}\right\}$, where the levels $\alpha_k^{(m)}$ are draw randomly (and this differently for each variable Φ_k). Thus, at each Monte Carlo iteration m, one obtains the lower & upper responses: $R_{LOW}^{(m)} = Inf\langle\mathcal{M}\left\{P_1^{(m)}, \ldots, P_N^{(m)}; A_1^\alpha, \ldots, A_K^\alpha\right\}\rangle; \ R_{UP}^{(m)} = Sup\langle\mathcal{M}\left\{P_1^{(m)}, \ldots, P_N^{(m)}; A_1^\alpha, \ldots, A_K^\alpha\right\}\rangle$

(5) Increment the Monte Carlo loop and return to Step (1) if $m < M$, else stop.

Fig. 4.4   (Plate): IRS algorithm for uncertainty propagation (Independent Random Sets)

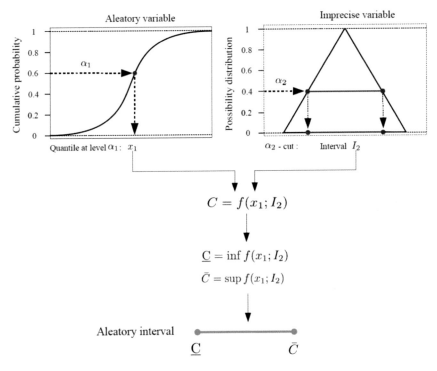

Fig. 4.5 Schematic of the IRS algorithm (Independent Random Sets) for propagating random + fuzzy (possibilistic) uncertainty. Here, the model response function is called "f", the model output is named "C" for "Concentration", and there are two uncertain parameters: X_1 (probabilistic), and X_2 (possibilistic), the latter represented by its α-cuts interval "I_2". Because of Monte Carlo sampling of the uniform random numbers α_1 and α_2, the lower and upper concentrations $\underline{C}$ and $\overline{C}$ are both random, characterized by lower & upper probability CDF's $F_{\underline{C}}^{LOW}(c)$ and $F_{\overline{C}}^{UP}(c)$

4.2.4 Comparative Review: Softwares and Uncertainty Propagation

This section presents a comparative review on software packages and some applications of fuzzy or probabilistic uncertainty analyses in the literature. Note first that several authors have developed comparisons of fuzzy-based approaches versus probabilistic Monte Carlo methods for uncertainty analyses (*some of these comparative works are mentioned below*).

Applications of fuzzy logic versus random Monte Carlo sampling

In the classical implementations of fuzzy logic, authors typically use triangular membership functions for each input parameter, and they obtain the output membership function through fuzzy logic operations of multiplication, addition, division, without employing repeated Monte Carlo simulations: see for instance [5]. In their work, based on fuzzy logic, Monte Carlo simulations were only used as an auxiliary tool for generating random conductivity fields and for comparing fuzzy logic results with hydraulic head moments obtained from Monte Carlo simulations.

Guyonnet et al. [12] applied a fuzzy "possibilistic" approach (without Monte Carlo), and compared it with probabilistic Monte Carlo simulations, for a very simple analytical model of vertical infiltration and dispersive transport solved explicitly in closed form. In their definition, fuzzy "possibilistic" analysis consists in propagating α-cuts through fuzzy logic operations, while probabilistic Monte Carlo involves multiple random sampling of input parameters leading to the constructed CDF/PDF of the output criterion. In their fuzzy logic implementation, they used triangular membership functions for each input parameter, and they obtained the output membership function through fuzzy logic operations (multiplication, etc.). Their output criterion was a measure of "excess cancer" (deduced from concentration and dose). In their final comparison, they graphically superimposed the membership function of excess cancer, and its PDF or frequency histogram constructed from the Monte Carlo simulations. The two results, as shown in the example of Fig. 4.6, are qualitatively similar, but differ significantly at large values; the probabilistic Monte Carlo approach underestimates the probability of excess cancer compared to the fuzzy logic result.

The conclusions of [12] is that fuzzy logic is "conservative," because it considers all possible combinations of uncertain parameter values, while in contrast, the probabilistic Monte Carlo approach under-represents parameter values having low probability of exceedance, because those have less chance of being randomly sampled

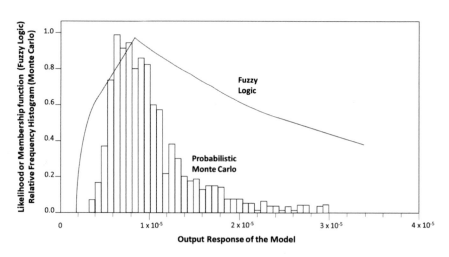

Fig. 4.6 Example comparison between probabilistic Monte Carlo results (empirical PDF of output response) and fuzzy logic results (membership function of output response) [After an example in [12]]

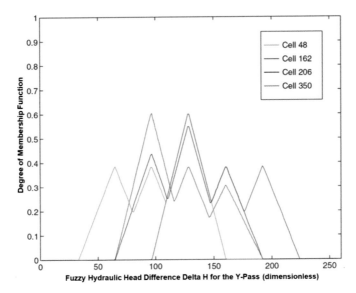

Fig. 4.7 Examples of fuzzy membership functions obtained in several 2D cells with the Fuzzy Logic code FLO2SIM, for the output variable Δh (hydraulic head difference) [Modified after [5]]

(*they might even not be sampled at all*). They conclude that Monte Carlo essentially eliminates worst case scenarios, while fuzzy logic preserves "low probability" scenarios, which is important for risk assessment.

Bagtzoglou et al. [5] also applied (non-possibilistic) fuzzy logic and arithmetic rules, to propagate uncertainty in two models: (1) a dedicated model of 1D unsaturated flow with fuzzy parameters, and (2) a fuzzy logic model of 2D saturated groundwater flow with fuzzy permeabilities (spatially homogeneous or not). The groundwater flow tests involved both synthetic problems, and a real test case from the WIPP waste disposal site (southeastern New Mexico, USA). The results of their dedicated Fuzzy Logic code (FLO2SIM) were analyzed in terms of membership functions of the output hydraulic heads in each numerical cell, as shown for example in Fig. 4.7. The output membership function was calculated according to the fuzzy logic and arithmetic rules summarized earlier in Sect. 4.1.2. The results of FLO2SIM were also compared with probabilistic Monte Carlo simulations performed with a conventional finite-difference code.

Remark. As can be seen from Fig. 4.7, membership functions of *fuzzy outputs* (like hydraulic head differences) are more complex than the triangular or trapezoïdal memberships of the *fuzzy inputs* (like permeabilities). Thus, *fuzzy output* functions $\mu(X)$ can have several maxima, and also, their highest value often remains less than 100%, that is Max$\{\mu\} < 1$, implying that all values of X are strictly fuzzy, none is "crisp."

Codes and packages on uncertainty propagation (summary overview)

In addition to FLO2SIM discussed above [5], several packages on uncertainty propagation have been devised since the 2000s and during the last decade. Uncertainty propagation capabilities have been implemented within specialized computer codes such as PEST and iTOUGH2 (to be reviewed further below). Uncertainty propagation packages are also available within open source software platforms like *OpenTURNS* and the *R-Project*:

- OpenTURNS software project for the *"Treatment of Uncertainties, Risks, aNd Statistics"* (https://openturns.github.io/www/), which is documented in [13], and
- Open source *R-Project for Statistical Computing* (https://www.r-project.org/).

For completeness, here are just a few examples of specific uncertainty packages related to uncertain propagation and risk analyses by various methods:

- HYRISK: "An R package for hybrid uncertainty analysis using probability, imprecise probability and possibility distributions," described in [11]. (See also discussion of the IRS algorithm in Sect. 4.2.3 comparing briefly the IRS with the "Hybrid approach": the purpose of both methods is to propagate probabilistic + fuzzy uncertainty).
- R-Project: [14]'s book documents several methods implemented within the "R" project, pertaining to probabilistic Uncertainty Quantification and Sensitivity Analyses for models with multiple input parameters, either directly or via metamodels.
- SmartUQ: Uncertainty Quantification package, incorporated in Comsol Multiphysics.
- UQlab (https://www.uqlab.com/): The Framework for Uncertainty Quantification: a Matlab ToolBox developed at the Chair of Risk, Safety and Uncertainty Quantification of ETH Zurich under the supervision of Prof. B. Sudret and Dr. S. Marelli.
- ESPER-1 package (*by the authors of this book*): this package is reviewed further below in Sect. 6.1. It was documented earlier in the ESPER public report by [15], under the name "ESPER software v.1.0". The report describes (i) the 3D semi-analytical model of concentration $C(x, y, z, t)$ and $Mass(t)$; and (ii): the probabilistic input/output Monte Carlo analysis of uncertainty propagation through this 3D model.

References

1. L.A. Zadeh, Outline of a new approach to the analysis of complex systems and decision processes. IEEE Trans. Syst. Man Cybern. SMC-**3**, 28–44 (1973)
2. L.A. Zadeh, Fuzzy sets as a basis for a theory of possibility. Fuzzy Sets Syst. **1**, 3–28 (1978) [originally: Memo UCB/ERL M77/12 University of California, Berkeley (1977)]

3. C. Baudrit, D. Dubois, D. Guyonnet, Joint propagation and exploitation of probabilistic and possibilistic information in risk assessment models. IEEE Trans. Fuzzy Syst. **14**, 593–608 (2006)

4. A. Barua, L.S. Mudunuri, O. Kosheleva, Why trapezoidal and triangular membership functions work so well: towards a theoretical explanation. J. Uncertain Syst. **8** (2014), www.jus.org.uk

5. A.C. Bagtzoglou, R. Ababou, A. Nedungadi, B. Sagar, Fuzzy rule-based hydrologic models for performance assessment of nuclear waste disposal sites. ASCE J. Hydrol. Eng. **14**(11), 1240–1248 (2009)

6. D. Guyonnet, Y. Ménard, C. Baudrit, D. Dubois, *HyRisk—Traitement des Incertitudes en Evaluation des Risques*. Rapport BRGM/RP 53714, September 2005, Rapport public (2005), 42pp.

7. D. Guyonnet, B. Bourgine, D. Dubois, H. Fargier, B. Côme, J. Chilès, Hybrid approach for addressing uncertainty in risk assessments. ASCE J. Environ. Eng. **129**(1), 68–78 (2003)

8. G. Coletti, D. Petturiti, B. Vantaggi, Fuzzy memberships as likelihood functions in a possibilistic framework. Int. J. Approx. Reason. **88**, 547–566 (2017). ISSN 0888–613X, https://doi.org/10.1016/j.ijar.2016.11.017

9. D. Dubois, D. Guyonnet, Risk-informed decision-making in the presence of epistemic uncertainty. Int. J. Gen. Syst. **40**(2), 145–167 (2011). https://doi.org/10.1080/03081079.2010.506179

10. C. Baudrit, D. Dubois, Comparing methods for joint objective and subjective uncertainty propagation with an example in a risk assessment, in *Paper in the 4th International Symposium on Imprecise Probabilities Application,* Pittsburgh, Pennsylvania (2005), 10p.

11. J. Rohmer, J.-C. Manceau, D. Guyonnet, F. Boulahya, D. Dubois, An R package for hybrid uncertainty analysis using probability, imprecise probability and possibility distributions (2018), 20p, https://eartharxiv.org/repository/view/1236/ (Non-peer reviewed paper preprint) [First author Institution: BRGM, France]

12. D. Guyonnet, B. Côme, P. Perrochet, A. Parriaux, Comparing two methods for adressing uncertainty in risk assessments. ASCE J. Environ. Eng. **125**(7) (1999)

13. M. Baudin, A. Dutfoy, B. Iooss, A.-L. Popelin, OpenTURNS: An industrial software for uncertainty quantification in simulation, in *Handbook of Uncertainty Quantification*, ed. by R. Ghanem, D. Higdon, H. Owhadi (Springer, 2017), 46p. HAL-01107849v2

14. S. Da Veiga, F. Gamboa, B. Iooss, C. Prieur, *Basics and Trends in Sensitivity Analysis (Theory and Practice in "R")*. SIAM—Society for Industrial and Applied Mathematics, Philadelphia PA (2021), xvi+291pp. https://doi.org/10.1137/1.9781611976694, https://epubs.siam.org/doi/abs/10.1137/1.9781611976694

15. J. Chastanet, J.-M. Côme, R. Ababou, M. Quintard, M. Marcoux, N. Tribouillard, *Project ESPER Evaluation of the sensitivity of prediction models for NAPL sources extinction and remediation: deterministic and probabilistic approaches to secure management decision (ESPER software Version 1.0—User's Guide)*. Public Report, ADEME, France, May 2019 (2019), 26pp. [in English]

Chapter 5
Applications of Uncertainty Analyses on Simplified Models

This chapter implements various simplified models of subsurface pollutant transport, and other related phenomena (e.g. geochemical corrosion of nuclear waste canisters). The goal is to illustrate specific applications of uncertainty analysis methods, probabilistic or fuzzy, reviewed earlier. Note: applications of uncertainty analyses for more complex 3D hydrogeological models of subsurface pollution are reserved for the next chapter (Chap. 6).

5.1 Introduction: Models with Uncertain Parameters (Classification)

We use the term "Input–Output" to emphasize the fact that models can be considered as operators $\mathcal{M}\{\dots\}$ relating input parameters to output variables:

$$Outputs = \mathcal{M}\{Inputs\} \tag{5.1}$$

The models can be implemented deterministically, with known inputs, or they can be implemented in a non-deterministic fashion for uncertainty analyses with imperfectly known parameters (random or fuzzy). In hydrogeology, most models are based on Partial Differential Equations (PDEs) or Ordinary Differential Equations (ODEs). One may distinguish, broadly:

(i) fully analytical models (not discretized);
(ii) quasi-analytical models (containing special functions or single integrals); and
(iii) numerically solved models based on spatial or space–time discretizations (Finite Volumes, Finite Elements, other Variational or Weighted Residual methods).

© The Author(s), under exclusive license to Springer Nature Singapore Pte Ltd. 2023
R. Ababou et al., *Uncertainty Analyses in Environmental Sciences and Hydrogeology*,
SpringerBriefs in Applied Sciences and Technology,
https://doi.org/10.1007/978-981-99-6241-9_5

Outputs can be expressed more or less explicitly in terms of inputs in analytical models (i) and (ii), but not in numerical models (iii). However, even with analytical models (i) or (ii), it may be difficult to express explicitly the *uncertainty distribution* of outputs in terms of the *uncertain* inputs. Explicit quantification of output uncertainty (e.g., via probability law) may be feasible only for the simplest analytical models. An example is the linear additive input/output models $Y = c_1 X_1 + c_2 X_2$. If both X_1 and X_2 are uniformly distributed, classical probability arguments lead to a triangular PDF of output Y. The second example is less trivial: it is the familiar temporal differential equation of first order kinetics. The integrated form of this model is $Y(t) = Y_0 \exp(-\lambda t)$. The decay constant λ is considered random, so the output $Y(t)$ is also random at any fixed time t. Depending on the probability law of λ, it may be feasible to express the probability law of $Y(t)$ analytically [1]. A fully analytical probabilistic study of $Y(t)$ is presented below for a uniform λ distribution (Sect. 5.2).

In the remainder of this section, we illustrate various methods of uncertainty analysis using simplified models of phenomena like temporal decay kinetics, growth of corrosion pits on a nuclear waste canister, or 1D (x, t) NAPL dissolution and solute transport in groundwater, all with uncertain parameters. More complex 3D hydrogeological models of subsurface contamination will be used later in the "applications" Chap. 6.

5.2   Decay Kinetics with Uncertain Decay Parameter

5.2.1   Introduction to the First Order Decay Kinetics Model

We consider here the temporal input/output differential model of first order decay

$$C_{RAND}(t) = C_0 \exp(-\lambda_{RAND}t) \qquad (5.2)$$

where C is solute concentration, λ is the decay constant (inverse time), and the subscript "$_{RAND}$" indicates a random (probabilistic) quantity. Thus, given the random input parameter λ_{RAND}, the output concentration $C_{RAND}(t)$ is also random for each fixed time t.

This model (Eq. 5.2) was previously analyzed probabilistically for several probability distributions of λ [1]. In their work, $C(t)$ represented a pressure difference and λ an *uncertain* matrix-fractures exchange parameter (fractured porous medium). The exchange parameter $\lambda > 0$ was taken random, and several probability laws were tested. The probability law of $C(t)$ was obtained explicitly without Monte Carlo simulations. Equation 5.2 can also serve as a model of fluid and heat exchange in matrix/fracture geothermal reservoirs, or radioactive decay with $C(t)$ standing for radionuclide mass, or 1rst order decay kinetics of solute concentration $C(t)$. Here we adopt the latter viewpoint.

5.2.2 *Exact Explicit Analytical Expression of C(t)'s Probability Law*

The method used by Kfoury et al. [1] can be used to express the probability law of the decaying concentration curve $C(t)$, as follows:

- First, note that for any fixed time t, the concentration C is a monotonic function of the random constant λ and of the deterministic time t. Also recall that the initial concentration C_0 is assumed to be deterministic. Rescaling $C(t)$, the model equation Eq. 5.2 can be written as $C_* = exp(-\Lambda_*)$, so that the input/output model is of the form:

$$\Lambda_* \rightarrow C_* = \exp(-\Lambda_*); \qquad C_* = C(t)/C_0; \Lambda_* = \lambda t \qquad (5.3)$$

  Note that Λ_* is the random constant $\Lambda_* = \lambda t$ for each fixed time t.
- The exact probability distribution (CDF or PDF) can now be obtained by using the fact that the input/output model $\Lambda_* \rightarrow C_* = \exp(-\Lambda_*)$ of Eq. 5.3 is monotonic. That is, the random output C_* is a monotonic decreasing function of the random input Λ_* for each fixed time t. The CDF of C_* can be obtained by probability arguments, or equivalently, the PDF of C_* can be obtained from the PDF of input Λ_* divided by the derived transformation $C_*(\Lambda_*)$, which is exponential.

Let us express the output PDF in terms of input PDF, using the exponential transform between input and output in this model. We obtain, from a well known probability result (e.g., [2]):

$$f_{C*}(c_*) = f_{\Lambda_*}(\lambda_*)/\left|g'(\lambda_*)\right|$$
$$\text{where} \quad g(\lambda_*) = \exp\{-\lambda_*\} \rightarrow \left|g'(\lambda_*)\right| = \exp\{-\lambda_*\}$$

Now, since we are interested in the probability law of C_*, we must express the inverse relation $g^{-1}(c_*) = \lambda_*(c_*)$ between input and output. This inverse relation is $\lambda_*(c_*) = -Ln\{c_*\}$, so the previous relation between the two PDFs becomes:

$$f_{C*}(c_*) = f_{\Lambda_*}(-Ln\{c_*\})/c_*$$

Finally, going back to unscaled quantities, re-introducing time t and initial concentration c_0, we obtain:

$$f_{C(t)}(c(t)) = f_\Lambda\left(-Ln\left\{\frac{c(t)}{c_0}\right\}\right)/(t \times c(t)) \qquad (5.4)$$

The mathematical expectation of $C(t)$, or its mean $\overline{C}(t)$, also denoted $m_C(t)$, is then deduced from its PDF with the change of variable $c \rightarrow b = -Ln\{c/c_0\}$; this yields the mean concentration curve:

$$m_C(t) = \overline{C}(t) = C_0 \int\limits_{-\infty}^{+\infty} f_\Lambda(b) \exp\{-bt\} db \qquad (5.5)$$

The concentration CDF $F_C(c(t)) = Proba\{C(t) \leq c(t)\}$ is deduced directly for each fixed time t from the monotonous transformation $\lambda \rightarrow C(t) = C_0 \exp(-\lambda t)$ by probability arguments, whence (renaming c(t) as C(t) in this final result):

$$F_C(C(t)) = 1 - F_\Lambda \left(-Ln \left\{ \frac{C(t)}{C_0} \right\} / t \right) \qquad (5.6)$$

Based on this result, the exact analytical probability law of $C(t)$ was then obtained *explicitly* for some distributions of the input λ, and the time-dependent moments of $C(t)$ could be deduced from its probability law.

5.2.3 The Mean Concentration Curve C(t)

For instance, focusing on the mean curve $\overline{C}(t)$, we provide here explicitly the exact analytical result for the case of uniformly distributed decay constant λ in the interval $[\lambda_{MIN}, \lambda_{MAX}]$:

- Mean Concentration: $m_C(t) = \overline{C}(t) = C_0 \frac{\exp\{-\lambda_{MIN}t\} - \exp\{-\lambda_{MAX}t\}}{(\lambda_{MAX} - \lambda_{MIN}) \times t}$

Furthermore, the concentration PDF given just above can also be used to infer the exact concentration variance and its square root, the standard deviation $\sigma_{C(t)}$, which is a measure of the *uncertainty* of concentration at each fixed time t *(this is left as an exercise for the interested reader)*.

Figure 5.1 shows the mean concentration curve $\overline{C}(t)$ for two probability laws of the random decay constant λ (uniformly and exponentially distributed, respectively), and this for various calculation methods: (1) the exact mean (the exact result given analytically), but also (2) the "naïve mean" where the random parameter λ is simply replaced by its mean $(\lambda_{MIN} + \lambda_{MAX})/2$ *(which is incorrect)*, and finally (3) an approximate mean obtained by a perturbation method *(not detailed here)*. The figure is adapted from [1].

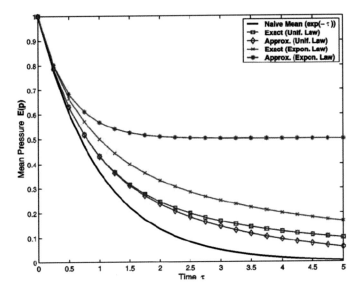

Fig. 5.1 Temporal relaxation of mean concentration $\overline{C}(t)$, for 2 probability laws of the random decay constant λ (uniform or exponential), and for various moment calculation methods: exact analytical, naïve mean (*incorrect*), and approximate perturbation. (Adapted from [1])

5.3 Taylor Expansion Around the Mean: MVFO Method and Application to Corrosion Pits

We now illustrate an *approximate* analytical method to evaluate the variability of a random output "$Z(t)$" using as an example a simple (but nonlinear) input/output model describing the growth of corrosion pits in a nuclear waste canister (to be explained below).

The method is usually named "MVFO" (Mean Value First Order); it consists essentially in a first order Taylor expansion of the output $Z(t)$ around its mean, which allows typically to calculate second order moments (variance) of the output, and also to calculate output probability law (if all the inputs are Gaussian, the approximate output obtained by MVFO is also Gaussian). This type of method based on Taylor expansion has several variants and bears several names. It is sometimes called "FOSM" for "First Order Second Moments." It is also known as *First Order Differential Analysis,* for sensitivity analyses as well as uncertainty propagation [3, 4]. The MVFO method can be extended to higher order Taylor expansions. In practice, MVFO is often used in reliability analyses (e.g., with more or less simplified models of structural failure, corrosion failure, and other probabilistic industrial or environmental events). It can also be used as a "fast" method for locating probability regions when running *importance sampling schemes* in Monte Carlo procedures.

Geochemical model of corrosion pit growth $Z(t)$ on a nuclear waste canister

Introductory review of the corrosion pits model in the literature

To illustrate the first order Taylor expansion approach, or "MVFO" method, we use the following empirical, non-differential geochemical model of corrosion pit growth $Z(t) = g(Parameters; t)$. The model output, corrosion pit depth $Z(t)$, is an algebraic function of both deterministic and uncertain parameters, as well as a function of time [5]. Physically, corrosion pits on the steel wall of nuclear waste canisters are due to oxidation and other mechanisms. We will implement in this section uncertainty analysis of this empirical model, distinguishing the different types of variables involved: the uncertain output $Z(t)$, the uncertain input parameters, the known deterministic input parameters, and the «time parameter». This problem is amenable to reliability analysis. For instance, "failure" can be defined as the event $Z(t) \geq Z_{CRIT}$, where Z_{CRIT} is critical pit depth (say 2 or 3 mm).

The MVFO method was initially adapted to corrosion pit modeling by the first author circa 2008, based on the corrosion pit model of Sutcliffe [5] and the unpublished report by Wu and Nair [6]. A similar corrosion pit model, also adapted from *Sutcliffe*, was analyzed with the MVFO method by Shih and Lin [7]. They used MVFO for sensitivity analyses, calculating the contribution of each uncertain parameter to the total variance of output $Z(t)$. In their work, the ratio of contributed variance of a given parameter to total variance yields a measure of sensitivity with respect to the given parameter (although the degree of sensitivity of each parameter is better "measured" using the root-mean-square norm, i.e., square root of contributed variance divided by square root of total variance).

The model of corrosion pit growth with uncertain parameters

The proposed probabilistic geochemical model predicts the depth $Z(t)$ of a corrosion pit as a function of time and other factors, for a spent fuel canister at 100 °C, as follows:

$$\underline{\underline{Z}}(t) = \underline{\underline{K_p}} \times K \times \exp\left\{\frac{a}{H}\right\} \times (O)^b \times (Cl)^c \times t^{\underline{\underline{n}}} \tag{5.7}$$

Table 5.1 defines the variables and parameters of this model. The first four quantities (underlined in Eq. 5.7) are random variables, namely, the uncertain output $Z(t)$ and three uncertain input parameters. Their probability laws are specified (Lognormal or Gaussian), as well as their mean "m" and standard deviation "σ". The remaining seven quantities are deterministic variables: they are perfectly known, including the time coordinate "t". Note that the time exponent "n" is itself a random parameter (positive and real valued), and that, remarkably, in this form of the model, the units of t^n and of a few other parameters depend on the random value of "n". Thus, in the original model of [5], $Z(t)$ has n—dependent units...which were expressed using the mean of "n" (instead of the random "n" itself). The first author of the present book has produced a rescaled version of this model by scaling the temporal term as $(t/t_{REF})^n$, such that temporal units do not depend on the random value of "n". However, for convenience, we continue here with the original unscaled model.

Table 5.1 Table of variables in the probabilistic model of corrosion pit depth Z(t)

$Z(t)$	Random…	Corrosion depth	mm	–
K_P	Random Lognormal	Pitting factor (empirical)	$mm/(year)^{n/2}$	$m = 4$; $\sigma = 1.0$
Cl	Random Gaussian	Chlorine concentration	$[\mu g/g]$ or [ppm]	$m = 6.5$; $\sigma = 0.65$
n	Random Lognormal	Empirical time exponent (real)	[dimensionless]	$m = 0.47$; $\sigma = 0.0329$
K	Deterministic	Corrosion factor (empirical)	$mm/(year)^{n/2}$	0.1706
O	Deterministic	Oxygen concentration	$[\mu g/g]$ or [ppm]	7
H	Deterministic	Absolute temperature	$[^{\circ}K]$	373
a	Deterministic	Temperature factor	$[^{\circ}K]$	-1402
b	Deterministic	Exponent (real-valued)	[dimensionless]	0.2
c	Deterministic	Exponent (real-valued)	[dimensionless]	0.543
t	Deterministic	Time coordinate	[years]	

Interpretation of the probabilistic pit depth Z(t) in the context of reliability analyses

Corrosion pit depth $Z(t)$ intervenes among several other mechanisms requiring more complex models of spent fuel container lifetime. In terms of probabilistic reliability analysis, the PDF of Time-to-Failure (TTF) of a container, or a system of containers, is related to the PDF of corrosion depth among other factors. The time horizon of interest may be 300 years, 1000 years, or thousands of years typically, but actually, the time horizon is a result of the calculation. For instance the question asked may be: what is the TTF (in years) if failure is defined as $Proba\{Z(t) > Z_{CRIT}\} = 0.01$? In reliability theory, the reliability function is defined as:

$$R(t) = Proba\{Z(t) \leq Z_{CRIT}\}. \tag{5.8}$$

The 99% TTF would then be solution of

$$R(t) = 0.990 \Rightarrow t_{99\%} \text{ (the 99\% Time-to-Failure)}.$$

In practice, the Time-to-Failure $t_{99\%}$ might be typically hundreds or thousands of years, with Z_{CRIT} on the order of millimeters (e.g. 2 or 3 mm). In order to obtain the TTF, the probability distribution of $Z(t)$, or at least the moments of $Z(t)$, must be known. This is the objective of the MVFO method based on Taylor expansion.

Moments and probability law of corrosion pit depth Z(t) by the MVFO method

First note that our corrosion model can be expressed as:

$$Z(t) = g(t; X_1, X_2, X_3)$$

where X_i are the input random variables, $Z(t)$ is the output random variable, and $g(\dots)$ is the nonlinear model function. In the case of the corrosion model above, the X_i variables are:

$$X_1 = K_P; \quad X_2 = Cl; \quad X_3 = n$$

It is difficult to compute analytically the exact probability law of $Z(t)$, due to non linearity of the model input/output function $g(\dots)$. The *MVFO* is an approximate method to obtain the probability law and moments of $Z(t)$ using a linearized version of the original random model. Linearization is obtained via a Taylor expansion of $g(X_1, X_2, X_3)$ around mean values:

$$Z(t) = g(t; X_1, X_2, X_3)$$

$$\approx g(t; m_1, m_2, m_3) + \sum_{i=1}^{i=3} \frac{\partial g}{\partial X_i}(t; m_1, m_2, m_3) \times (X_i - m_i) \qquad (5.9)$$

...or:

$$Z(t) \approx a_0(t) + \sum_{i=1}^{i=3} a_i(t) \times (X_i - m_i) \text{ with}: \ a_0(t) = g(t; m_1, m_2, m_3);$$

$$\text{and}: \ a_i(t) = \frac{\partial g}{\partial X_i}(t; m_1, m_2, m_3).$$

The $a_i(t)$ coefficients can be viewed as *"mean sensitivities"*. For the corrosion model, they are:

$$a_0(t) \approx a_0 \times (t)^{0.47}; a_1(t) \approx a_1 \times (t)^{0.47};$$
$$a_2(t) \approx a_2 \times (t)^{0.47}; a_3(t) \approx a_0 \times \ln(t) \times (t)^{0.47}.$$

To sum up, at this point, we have obtained for any fixed time t, an explicit expression of the random $Z(t)$ as a linear combination of random input parameters $\{X_1, X_2, X_3\}$. The mean and variance of $Z(t)$ are easily computed from above relations, $\{X_1, X_2, X_3\}$ being uncorrelated. Thus, the mean of $Z(t)$ is simply given by the 0-order coefficient:

$$E\{Z(t)\} \equiv \overline{Z}(t) \approx a_0(t) = \alpha_0 \times (t)^{0.47} \text{ (here in millimeters)}.$$

...and the variance of $Z(t)$ is given by:

$$Var\{Z(t)\} \equiv \sigma_Z^2(t) \approx \sum_{i=1}^{i=3} a_i^2(t)\sigma_{Xi}^2$$

$$= \left\{\alpha_1^2\sigma_{X1}^2 + \alpha_2^2\sigma_{X2}^2 + \alpha_3^2\sigma_{X3}^2[Ln(t)]^2\right\} \times (t)^{2m_{X3}}$$

…where $X_1 = K_P$; $X_2 = Cl$; $X_3 = n$. For instance $m_{X3} = 0.47$ is the mean value of parameter n. Taking the square root of $\sigma_Z^2(t)$ yields the *root-mean-square measure of uncertainty* $\sigma_Z(t)$.

It is less straightforward to determine the probability distribution of $Z(t)$. Recall that $\{X_1, X_2, X_3\}$ are respectively Lognormal, Gaussian, and Log-normal. The resulting linear combination $Z(t)$ is a priori neither Gaussian nor Lognormal. This complicates the task of determining probabilities such as Time-to-Failure of $Z(t)$ in the context of reliability analysis.

We circumvent this problem, below, by accepting a Gaussian approximation of $Z(t)$.

Time-to-Failure for the corrosion pit depth model $Z(t)$ from the MVFO method

To simplify calculations of the probability distribution of $Z(t)$, we admit a Gaussian approximation for the two Log-normal variables $X_1 = K_P$ and $X_3 = n$. This is admissible because their coefficient of variation is much less than one, as can be seen in the right column of Table 5.1 (see moments relations earlier in Sect. 3.1.1). With this approximation, all three input parameters $\{X_1, X_2, X_3\}$ are now Gaussian and independent, $N(m_{Xi}, \sigma_{Xi}^2)$, and it is now easier to compute the probability law of $Z(t)$ as a linear combination of Gaussian's with known means (m_{Xi}) and variances (σ_{Xi}^2). The resulting $Z(t)$ is itself Gaussian:

$$Z(t) = N\left(m_Z(t), \sigma_Z^2(t)\right)$$

with: $m_Z(t) \approx a_0(t) = \alpha_0 \times (t)^{0.47}$ and: $\sigma_Z^2(t) \approx \sum_{i=1}^{i=3} a_i^2(t)\sigma_{Xi}^2$ (see just above).

The Gaussian CDF of $Z(t)$, that is $F_{Z(t)}^{GAUSS}(z) \equiv Proba\{Z(t) \leq z\}$, can now be calculated analytically using the $erf(\dots)$ function, as explained in the probability Sect. 3.1. We provide below some results deduced from this analytical characterization of $Z(t)$: (1) evaluation of the 99% Time-to-Failure $t_{99\%}$, and (2) plot of the reliability function $R(t)$.

(1) *Evaluation of the 99% Time-To-Failure.*
With the linearized First Order and Gaussian approximation of $Z(t)$, the required probability relation to evaluate the 99% Time-to-Failure $(t_{99\%})$ is:

$$F_{Z(t_{99\%})}^{GAUSS}(Z_{CRIT}) = Proba\{Z(t_{99\%}) \leq Z_{CRIT}\} = 0.990 = R(t_{99\%}).$$

Since the CDF $F_{Z(t)}^{GAUSS}$ involves the error function $erf(\dots)$, inverting this relation to obtain $t_{99\%}$ requires the inverse error function $erfinv(\dots)$. With the available data on the moments of uncertain input parameters (Table 5.1), and given their approximation as Gaussian input variables, we obtain after

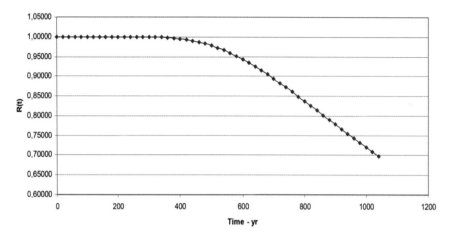

Fig. 5.2 Evolution of the reliability function $R(t)$ versus time in years, for a critical corrosion pit depth $Z_{CRIT} \approx 2$ mm. Note that $R(t) = 0.990 = 99\%$ at time $t_{99\%} \approx 440$ years (see text)

calculation:

$$\text{Time-to-Failure for } Z_{CRIT} \approx 2 \text{ mm}: \quad t_{99\%} \approx 440 \text{ years.}$$

At that time, $t \approx 440$ years, the previous moment calculations indicate that the standard deviation of $Z(t)$ is about 0.37 mm (measure of uncertainty of $Z(t)$).

(2) **Reliability function R(t).** Figure 5.2 shows a plot of the reliability function $R(t)$ defined in Eq. 5.8, taking $Z_{CRIT} \approx 2$ mm.

5.4 Coupled 1D Model of Contaminant Migration and Source Dissolution

The prototype 1D transport model used in this section is an advection–dispersion model for solute concentration $C(x, t)$, coupled to "oil" saturation $S(x, t)$ in the presence of a dissolving trapped NAPL source (DNAPL in this case). With constant porosity, water velocity, and dispersion coefficient, all independent of space and time, the model takes the form:

$$\begin{cases} \frac{\partial}{\partial t}\{\Phi(1 - S)C\} + \frac{\partial}{\partial x}\{VC\} - \frac{\partial}{\partial x}\{\Phi(1 - S)D\frac{\partial C}{\partial x}\} = \\ \qquad\qquad -\alpha_* \, \Phi(1 - S) \{C - C_{EQ}\} \\ \frac{\partial}{\partial t}\{\rho_{NAPL}\Phi S\} = +\alpha_* \, \Phi(1 - S) \{C - C_{EQ}\} \end{cases} \tag{5.10}$$

Remarks:

- The first equation in Eq. 5.10 is a balance equation for solute mass per m^3 of porous medium.
- The second of Eq. 5.10 is a balance equation for the trapped NAPL mass per m^3 of porous medium.
- The linear dissolution model with constant α_* is an approximation; α_* should in fact decrease with saturation; letting $\alpha_* = 0$ when saturation reaches some low residual value can improve the model: this was implemented in the application by letting $\alpha_* = 0$ for $S < S_{RESIDUAL} = 0.01$ (a smaller value could also be used).
- For our applications of uncertainty analyses, the 1D model defined by Eq. 5.10 will be *essentially* reduced to purely advective transport without dispersion (take $D = 0$ in the first equation). This advective model retains the coupling with NAPL dissolution via the exchange coefficient.

The two space–time variables of the model are:

- $C(x, t)$: solute concentration, $\frac{kg}{m^3 water}$ or $\frac{g}{m^3 water}$ (NAPL constituent dissolved in the water phase)
- $S(x, t)$: NAPL phase saturation (also named S_O for "oil saturation") in $\left[\frac{m^3 NAPL}{m^3 Pores}\right]$

The parameters of the model may be either deterministic or uncertain; they are:

- Φ : aquifer porosity in m^3/m^3 (also denoted ε)
- V: Darcy velocity in m/s or m/day (*NB: pore water velocity is* $V_{PORE} = V/\Phi$)
- D: dispersion coefficient in [m^2/s] or [m^2/day]; it can also be interpreted as a 1D diffusion–dispersion coefficient $D = D_0 + a.V$ where $a[m]$ is a longitudinal dispersivity length scale.
- ρ_{NAPL}: density of the trapped NAPL, in $\left[kg\, NAPL/m^3\, NAPL\right]$.
- C_{EQ}: equilibrium concentration of the solute (concentration at which NAPL dissolution stops)
- α_*: mass exchange coefficient [day^{-1}] in the NAPL dissolution process, to be understood as the exchange coefficient in $\left(\frac{Kg\, Solute}{m^3\, domain}\right)/\left(\frac{Kg\, NAPL}{m^3\, domain}\right)$ per day.

Initial conditions for the 1D *advective* transport problem, with $D \approx 0$, will be typically as follows (*boundary conditions are not needed for advective transport*):

> **Domain** : Semi-infinite domain $x \geq 0$ with a NAPL Source zone within
> $x \in [0, L_S]$, taking for instance $L_S = 1$
> **Initial Conditions** : $C(x, 0) \approx 0$ or C_{EQ}, and: (5.11)
> $S(x, 0) \approx 0.10$ for $x \in [0, L_S]$ (source zone);
> $C(x, 0) = S(x, 0) = 0$ for $x > L_S$ (outside the source).

Finally, in order to simplify the forthcoming uncertainty analyses, we simplify the formulation of Eq. 5.10 by capturing the exchange coefficient in the new form:

$$\alpha = \alpha_* \Phi(1 - S)$$

The meaning of the two mass balance equations in Eq. 5.10 remains unchanged. It appears that the new coefficient α depends on saturation, but if $S < 0.10$, say, then the new exchange coefficient α does not vary much with saturation during the dissolution/migration process. On the other hand, as stated before, we apply a threshold $\alpha = 0$ when saturation reaches some low residual value $S_{RESIDUAL}$.

Typical reference values of the parameters are:

- Darcy velocity: $V \approx 0.5$ to 1.0 m/day
- Porosity: $\theta \approx 0.04$
- Dispersion coefficient: $D \approx 2.5$ m/day^2 or... $D = 0$ here
- Exchange coefficient: $\alpha \approx 0.01$ day^{-1} (dissolution time scale $1/\alpha \approx 100$ days)
- NAPL density: $\rho_{NAPL} \approx 1.6 \ 10^{+6}$ g/m^3 (this is a Dense NAPL)
- Equilibrium concentration: $C_{EQ} \approx 150$ g/m^3
- Typical initial saturation: $S_{INIT} \approx 0.10$

5.4.1 Monte Carlo Analysis of 1D Dissolution/Transport via a Metamodel

We now present an example application of probabilistic uncertainty analysis based on the *metamodel* approach for the 1D concentration-saturation transport defined by Eqs. 5.10 and 5.11, where we consider here for simplicity the case $D = 0$ (results with or without dispersion were indistinguishable given the parameters values). The advective transport/dissolution model was solved with a classical discretization scheme using COMSOL Multiphysics, and the metamodel approach was implemented with the OpenTURNS software (https://openturns.github.io/www/; [8]).

Deterministic solution

Before proceeding with uncertainty analyses, the model solution with deterministic parameters should be examined briefly (see typical values in the previous subsection). The following quantities are of interest: average saturation over the source zone $\overline{S}(t)$ (which is directly related to the remaining NAPL mass in the source zone); and the concentration $C(x_{SPOT}, t)$ at a given point $x = x_{SPOT}$ downstream the source zone. These two quantities are represented in Fig. 5.3.

In Fig. 5.3, the concentration $C(x_{SPOT})$ increases suddenly then remains constant till about 400 days. The subsequent decrease of concentration $C(x_{SPOT})$ after 400 days is due to the dissolution of the source. Note that $\overline{S}$ has decreased to less than 20% of its initial value at 400 days. Eventually the source will be entirely spent.

Output "criteria" of interest

Output criteria should be defined for purposes of uncertainty analyses. For this 1D dissolution/transport problem, the following output criteria are of interest:

(1) Source depletion time $t_{DEPLETION}$ such that $\overline{S}(t_{DEPLETION}) = S_{DEPLETION}$. We define here $t_{DEPLETION}$ as the time taken to deplete the NAPL source to 1%

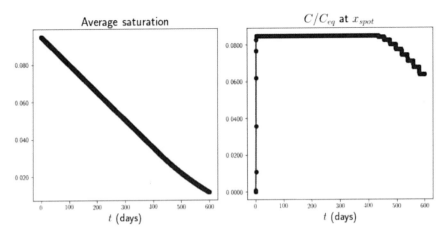

Fig. 5.3 Time evolution of $\overline{S}(t)$ and $C(x_{SPOT}, t)$ for deterministic parameters. Here $\overline{S}(t)$ is the average saturation over the source zone, related to the remaining NAPL mass in the source zone, and $C(x_{SPOT}, t)$ is solute concentration at a point x_{SPOT} downstream the source zone

of its initial value. With initial saturation in the source zone equal to 0.10, for instance, this would yield $S_{DEPLETION} = 0.001$. Note this is only a selected output criterion, not an input parameter of the model.

(2) Maximum concentration C_{MAX} corresponding to the plateau of $C(x_{SPOT}, t)$ visible in Fig. 5.3.

(3) Critical time t_{CRIT} such that $C(x_{SPOT}, t_{CRIT}) = C_{CRIT}$ where C_{CRIT} is a small critical concentration (for instance $C_{CRIT} = 0.1\,g/cm^3$).

The first criterion $t_{DEPLETION}$ provides a measure of the source lifetime. It is useful to calculate its lower bound or minimal value, by considering the case where dissolution occurs entirely at local equilibrium, that is instantaneously ($\alpha \to \infty$). The corresponding dissolution front velocity V_{FRONT} is then maximal; it is given by:

$$V_{FRONT} = VC_{EQ}/(\rho_{NAPL}\Phi S_{INIT})$$

and the minimal depletion time is given by:

$$t_{DEPLETION}^{MIN} = (L_{SOURCE}/V_{FRONT}) \times (1 - S_{DEPLETION}/S_{INIT}).$$

Metamodel construction: sampling random inputs and building a response function

As explained earlier (Sect. 2.2.2), the *metamodel approach* consists first in constructing a response function for a given output criterion, this being accomplished by sampling the uncertain inputs and running the model repeatedly (Monte Carlo).

Then, finally, the resulting response function is used instead of the model itself to propagate uncertainty from input parameters to the output criterion.

In the present case, our 1D dissolution/transport model is numerical; its ouputs are the space–time discretized solutions $C(x, t)$ and $S(x, t)$, from which the output criteria defined above can be computed. We will choose here two uncertain input parameters, the Darcy velocity V and the exchange coefficient α. A parsimonious sampling plan (Design of Experiments) called "Quasi Random Sequences" or "Low Discrepancy Sequences" was selected in OpenTURNS for sampling the random $\{V, \alpha\}$ parameters: it is shown in Fig. 5.4. The vertical and horizontal profiles indicate the density of points in each direction of the $\{V, \alpha\}$ plane. It can be seen that the density of points is fairly uniform in each direction. Note that other parsimonious sampling schemes are available, like LHS (Latin Hypercube Sampling), which was reviewed earlier in Sect. 2.3.2.

An output criterion is then defined, for example, the source depletion time $t_{DEPLETION}$ defined above. Given this output criterion, the response function to be constructed is therefore:

$$t_{DEPLETION} = f(V, \alpha)$$

Finally, a family of response functions $\widehat{f}(\ldots)$ must be selected to approximate the response of the model. Our choice for $\widehat{f}(\ldots)$ was a second order *Polynomial Chaos* function of 2 variables, available in OpenTURNS. The discrete response surface is constructed by computing the output criterion for each discrete sample of $\{V, \alpha\}$. The second order Polynomial Chaos surface is then fitted by nonlinear regression to

Fig. 5.4 Sampling of the two uncertain variables $\{V, \alpha\}$ with V as abscissa and α as ordinate. The variables are sampled in OpenTURNS according to a parsimonious sampling plan ("Quasi Random Sequences"). The vertical and horizontal histogram profiles (light blue shading) indicate the density of points in each direction in the $\{V, \alpha\}$ plane. *NB: the parameter plane is normalized here to* [0.0, 1.0] × [0.0, 1.0]

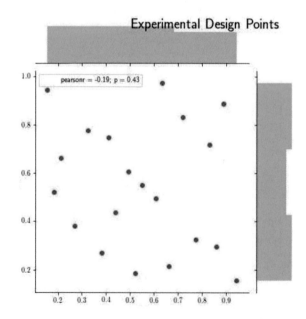

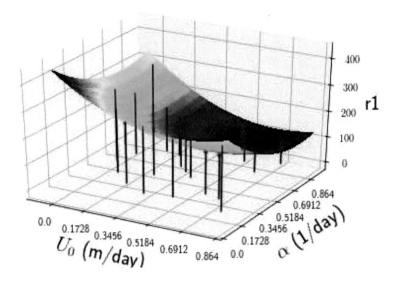

Fig. 5.5 Metamodel response surface for the output "$t_{DEPLETION}$" (denoted here "$r1$", in days), for two uncertain parameters, Darcy velocity (m/day) and exchange coefficient (1/day). The response surface (colored) was obtained by nonlinear regression fit of a second order Polynomial Chaos to the sample points $\{t_{DEPLETION}, V, \alpha\}^{(j)}$ shown as vertical bars

the points $\{t_{DEPLETION}, V, \alpha\}^{(j)}$, as shown in Fig. 5.5. The criterion $t_{DEPLETION}$ is denoted «$r1$» in the figure. The nonlinear regression coefficient was 0.97616, which is a fair fit.

Uncertainty propagation via the metamodel

Finally, the constructed metamodel was used for probabilistic Monte Carlo analyses of the uncertain ouput criterion $t_{DEPLETION}$ [denoted «$r1$» in the figures].

The effect of uncertain Darcy velocity V (denoted "U_0" in the figures) is first examined for a fixed deterministic $\alpha = 1.74 \ 10^{-6} \, \text{s}^{-1} = 0.150336 \, \text{days}^{-1}$, that is $1/\alpha \approx 6.65$ days. The metamodel response function was then sampled repeatedly for $N = 1000$ values of the uncertain velocity V ("U_0"), assuming a Gaussian velocity distribution.

Figure 5.6 shows the PDF of the input velocity and the PDF of the resulting output $t_{DEPLETION}$ [«$r1$»], which was obtained by sampling the *metamodel*'s response function. The output result is shown as a PDF histogram, together with a smoothed version of it. Note that the PDF of $t_{DEPLETION}$ [«$r1$»] is clearly non-Gaussian, positively skewed, with a somewhat "fat" tail toward large values of the depletion time. The most probable depletion time (mode of the PDF) is about 354 days, and the highest sampled values are less than 388 days.

Similarly, the effect of uncertain exchange coefficient α was then analyzed, for a Gaussian distribution of α, keeping the Darcy velocity fixed. The results were qualitatively similar: the resulting PDF of the random output $t_{DEPLETION}$ [denoted

Fig. 5.6 *Top*: Gaussian PDF of the random input parameter denoted "U_0" (Darcy velocity V). *Bottom*: empirical PDF of the random output criterion denoted «$r1$» (depletion time $t_{DEPLETION}$ in days), obtained by Monte Carlo sampling of the *metamodel* response function

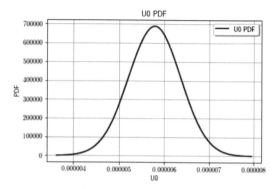

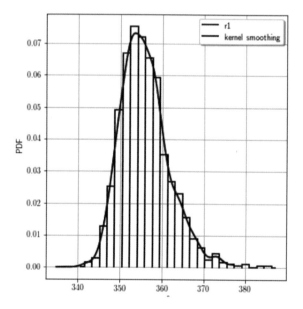

«$r1$»] obtained by sampling the metamodel's response function, showed that the PDF histogram of $t_{DEPLETION}$ [denoted «$r1$»] was non-Gaussian, positively skewed, with a "fat" tail towards large values of the depletion time. The most probable depletion time (mode of the PDF) was about 157 days and the highest sampled values are about 205 days.

5.4.2 Possibilistic Analysis of 1D Contaminant Migration with IRS

Here, we use a 1D transport model similar to the one described above (Eq. 5.10 without dispersion) to illustrate another technique of uncertainty propagation: the IRS method (Independent Random Sets), based on the "possibilistic" approach combining fuzziness and probabilistic uncertainty. This approach specifies a "possibilistic" distribution for each uncertain parameter, combining fuzzy membership function and probabilistic CDF, as explained earlier in Sect. 4.2.

The goal here is to show the type of result that can be obtained this way, without detailing much the underlying model. Briefly, the solute concentration is governed by a 1D space–time transport model coupled to NAPL dissolution, as in Eq. 5.10. In the uncertainty analyses below, $C(t)$ will represent the concentration output $C(x_1; t)$ at a given point in space ($x_1 = 5$ m).

Figure 5.7 shows two examples of input/output uncertainty propagation for the 1D concentration transport model, based on the theory of possibilities combining fuzzy variables and probabilities (CDFs). The ordinate axis shows the resulting CDF (Cumulated Distribution Function) of concentration $C(t_{40 \, days})$ at the fixed time $t = 40$ days. It is a probability of non-exceedance: $CDF(c) = Proba\{C \leq c\}$. The abscissa represents concentration $c = C(t_{40 \, days})$, or a normalized version of it.

In Fig. 5.7, in each sub-figure, several CDFs are shown because the output concentration is both probabilistic and fuzzy. In each case, four different CDFs are being displayed, corresponding to four confidence levels: $\alpha = 100\%$ for the *upper bound* CDF at left (red color); $\alpha = 66\%$ for the grey CDF; $\alpha = 33\%$ for the green CDF; and $\alpha = 0$ for the *lower bound* CDF at right (dark blue color). The two sub-figures differ in terms of the uncertain inputs:

- In the top sub-figure, the transport velocity V is a possibilistic parameter represented by a trapezoïdal membership function, or more accurately, a trapezoïdal possibility distribution $\mu_V\{0.5, 0.6, 0.8, 1.0\}$, also characterized by its associated upper and lower bound CDFs (as explained earlier in Sect. 4.2). This is a case of so-called "pure epistemic uncertainty."
- In the bottom sub-figure, velocity V is a log-normal probabilistic parameter, that is a random variable, having a fuzzy probability distribution (fuzzy moments). The mean velocity m_V, and its standard deviation σ_V, are both fuzzy within intervals $[m_{V0} \pm 30\% \, m_{V0}]$ and $[\sigma_{V0} \pm 30\% \, \sigma_{V0}]$. This is a case of so-called "imprecise aleatory uncertainty."

The latter case (bottom sub-figure) illustrates one way in which uncertainty can arise from a combination of randomness and fuzziness. Here, concentration (which is the output of the model) has a probabilistic uncertainty because the input parameter V is random (Log-normal), but has also uncertainty in terms of its fuzziness because the moments of the Log-Normal velocity V are not known precisely (they are not "crisp" but fuzzy, or "imprecise").

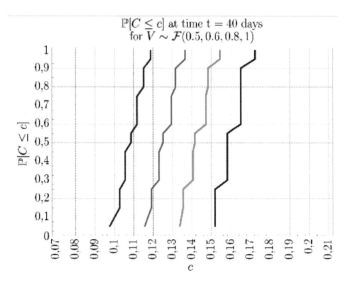

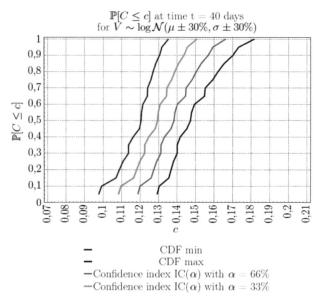

Fig. 5.7 Applications of the theory of "possibilities" combining fuzzy variables and probabilities (CDFs), implemented with the IRS technique. Solute concentration is governed by a simplified 1D advective transport model with uncertain velocity V. Here $C(t)$ represents $C(x_1; t)$ at a given point in space. *Top figure*: input velocity V is a "possibilistic" variable with trapezoidal membership function (and its associated CDFs). *Bottom figure*: velocity V is a log-normal random variable with fuzzy moments. In both cases the ordinate axis shows the resulting CDF of concentration $C(t_{40}$ days$)$. Four CDFs are displayed, corresponding to confidence levels $\alpha = 100\%$ (*upper bound* CDF at left in red color); $\alpha = 66\%$ (grey CDF); $\alpha = 33\%$ (green CDF); and $\alpha = 0$ (*lower bound* CDF at right in dark blue). The abscissa represents the concentration variable $c = C(t_{40}$ days$)$, normalized

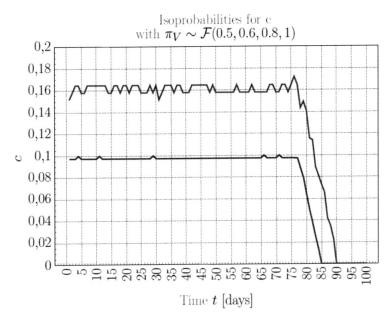

Fig. 5.8 Temporal evolution of two isoprobability concentrations $c(t)$, satisfying $Proba(C(x, t) \leq c(t)) = P_{MIN} = 0.5$ (lower blue curve $c(t)$) and $Proba(C(x, t) \leq c(t)) = P_{MAX} = 1.0$ (upper red curve $c(t)$), in the case of a possibilistic input parameter V with trapezoidal membership function $\mu_V\{0.5, 0.6, 0.8, 1.0\}$.

Other types of results can be obtained in this possibilistic framework. Thus, for the first case above, with possibilistic velocity parameter V having "trapezoidal distribution" $\mu_V\{0.5, 0.6, 0.8, 1.0\}$, we show in Fig. 5.8 the time evolution of two iso-probability concentrations $c(t)$, where $c(t)$ satisfies, respectively:

$$Proba(C(x_1, t) \leq c(t)) = P_{MIN} = 0.5 \,;$$
$$Proba(C(x_1, t) \leq c(t)) = P_{MAX} = 1 \,; \quad (both\ at\ x_1 = 5\ m).$$

Note, as expected, that the two iso-probability concentrations $c(t)$ in Fig. 5.8 drop to zero as the concentration front arrives and passes beyond the fixed position x_1.

References

1. M. Kfoury, R. Ababou, B. Noetinger, M. Quintard, Matrix-fracture exchange in a fractured porous medium: stochastic upscaling. (Coefficient d'échange matrice-fracture en milieu poreux fracturé : changement d'échelle par approche stochastique), in *C.R.A.S, Comptes Rendus Académie Sciences*, vol. 332 (C.R. Mécanique, Paris, 2004), pp. 679–686. Elsevier
2. A. Papoulis, S. Unnikrishna Pillai, *Probability, Random Variables, and Stochastic Processes*, 16 chapters, vol. 852, 4th edn.(Mc-Graw Hill, 2002)

3. D.A. Zimmerman, K.K. Wahl, A.L. Gutjahr, P.A. Davis, *A review of techniques for propagating data and parameter uncertainties in high-level radioactive waste repository performance waste repository performance assessment models.* NUREG/CR-5393 (SAND89–1432), US Nuclear Regulatory Commission(1989).
4. G.J. Hahn, S.S. Shapiro, *Statistical Models in Engineering* (Wiley, New York, NY, 1967), p.1967
5. W.G. Sutcliffe, Uncertainty analysis: An illustration from nuclear waste package development. Nucl. Chem. Waste Manage. **5**(2), 131–140 (1984)
6. Y. Wu, P. Nair, *Fast Probabilistic Performance Assessment (FPPA) methodology evaluation.* CNWRA 88–004, US.NRC Contract NRC-02-88-005, San Antonio TX, 42 p. [Unpublished report] (1988)
7. D.C.-F. Shih, G.-F. Lin, Uncertainty and importance assessment using differential analysis: an illustration of corrosion depth of spent nuclear fuel canister. Stoch. Environ. Res. Risk Assess. (SERRA) **2006**(20), 291–295 (2006). https://doi.org/10.1007/s00477-005-0028-z
8. M. Baudin, A. Dutfoy, B. Iooss, A.-L. Popelin, OpenTURNS: An industrial software for uncertainty quantification in simulation, in *Handbook of Uncertainty Quantification*, ed. By R. Ghanem, D. Higdon, H. Owhadi (Springer, 2017). HAL-01107849v2, 46 p.

Chapter 6
Applications of Uncertainty Analysis to 3D Subsurface Contamination Problems

In this section, we present two applications of uncertainty propagation for more complex 3D subsurface contamination problems, the first one modeled quasi-analytically with special functions and single integrals, and the second one modeled numerically with a complex computer code.

(1) *Model 1*. 3D transient advection–diffusion-dispersion of a contaminant, migrating in an aquifer, from a decaying rectangular patch source of NAPL. This mathematical model is quasi-analytical; it was obtained by us by extending previous solutions. It may be of interest not only for uncertainty analyses, as we do here, but also for benchmark tests in contaminant migration modeling.

(2) *Model 2*. 3D flow and transient advection–dispersion coupled to NAPL source dissolution, occurring in an actual polluted site. This problem is modeled numerically with the commercial MODFLOW-SURFACT™ code.

To sum up, the models used in this section are both 3D, and they are more complex than those already used for illustration in previous sections of this work.

The uncertainty analysis methods employed in this section are probabilistic but they differ technically for Model 1 (quasi-analytical) and Model 2 (numerical code) as follows:

(1) Monte Carlo simulations are applied directly to Model 1, with random input parameters taken one-by-one; whereas…

(2) Monte Carlo simulations are performed only indirectly in Model 2, using a pre-constructed metamodel approximation of the original model; this is justified by the costly computational nature of Model 2 (compared to the quasi-analytical Model 1).

6.1 Application of Uncertainty Analysis to 3D Contaminant Advection–Dispersion from a Decaying Source (Quasi-Analytical ESPER-1 Model)

We present here an application of uncertainty propagation for subsurface contaminant transport using a quasi-analytical model of 3D advection–dispersion of a dissolved contaminant, migrating as a concentration plume in an aquifer, and originating from a decaying rectangular patch source of trapped DNAPL (Dense Non Aqueous Phase Liquid). The source dissolution mechanism is simplified, but the migration process itself accounts for 3D advection, diffusion, and dispersion (with transverse and longitudinal dispersivities), retardation due to adsorption, and first order decay.

Before proceeding with uncertainty analyses, we describe first the 3D semi-analytical contaminant transport model equations and solution., which has not been published before except for a report by Chastanet et al. [1]. This model is an enhanced version of previous semi-analytical models. Uncertainty propagation analyses based on probabilistic Monte Carlo simulations with random input parameters will then be described in further subsections.

6.1.1 Model Equations: PDE, Geometry, Boundary and Initial Conditions

The proposed quasi-analytical model describes the 3D migration of solute concentration $C(x, y, z, t)$ by advection and dispersion, and the corresponding decaying mass of the dissolving NAPL source $M(t)$ which feeds the dissolved contaminant plume. Our solution is essentially a modified and extended version of the previous 3D solution of [2], which was itself preceded by the 3D solution of [3], and by another approximate 3D solution initially proposed by Domenico [4]. The [4] solution is fast to compute, but can produce errors, which were analyzed by West et al. [5], such as 80% underprediction of concentration along the centerline of the plume for some cases. Compared to these solutions, our 3D solution remains relatively simple (*see below*) and does not make any approximations. Furthermore, unlike its predecessors, our solution is applicable generally to a decaying NAPL source of finite duration (rather than infinite duration in the previous solutions). To be fair, there are other quasi-analytical solutions for advective–dispersive transport, some of them computationally complex, such as that of [6] which relies on Laplace Transforms, and incorporates a broad set of time-dependent boundary conditions (but not specifically a decaying NAPL source). Our solution does not involve Laplace Transform, and remains relatively simple to compute. It was implemented under the name "*Modèle 1.1 ESPER*" in the ESPER 1 package for uncertainty analyses [1]. This 3D solution, described below, may also be of interest for modelers running benchmark tests in contaminant hydrogeology.

Equational model: governing PDE

The 3D advective–dispersive migration of the concentration plume in groundwater is governed by the following constant coefficients PDE:

$$R\frac{\partial C}{\partial t} + V_X\frac{\partial C}{\partial x} = D_{XX}\frac{\partial^2 C}{\partial x^2} + D_{YY}\frac{\partial^2 C}{\partial y^2} + D_{ZZ}\frac{\partial^2 C}{\partial z^2} - R\lambda C \qquad (6.1)$$

The groundwater pore velocity V_X, the diffusion/dispersion coefficients D_{XX}, D_{YY}, D_{ZZ}, the retardation coefficient R, and the first order decay coefficient λ, are all assumed constant in space–time. This PDE can also be expressed using Einstein's rule, of implicit summation on repeated indices:

$$R\frac{\partial C}{\partial t} + V_1\frac{\partial C}{\partial x_1} = V_1\alpha_i\frac{\partial^2 C}{\partial x_i\partial x_i} - R\lambda C \qquad (6.2)$$

...where the pore velocity vector is aligned with axis Ox_1:

$$\vec{V} = \begin{bmatrix} V_X \\ 0 \\ 0 \end{bmatrix} = \begin{bmatrix} V_1 \\ 0 \\ 0 \end{bmatrix} \Leftrightarrow V_i = V_1 \times \delta_{1i} \qquad (6.3)$$

and the diagonal tensor coefficient D_{ij} incorporates both local isotropic diffusion $D_0\delta_{ij}$ and diagonal anisotropic dispersion as follows:

$$D_{ij} = D_0\delta_{ij} + \alpha_i V_1\delta_{ij} \qquad (6.4)$$

Three dispersivity length scales intervene here: the longitudinal dispersivity α_1 and the transverse dispersivities $\alpha_2 = \alpha_3$, all in meters. Coefficients $(\alpha_1, \alpha_2, \alpha_3)$ are also named $(\alpha_X, \alpha_Y, \alpha_Z)$, or $(\alpha_L, \alpha_T, \alpha_T)$. Accordingly, the 3 diagonal coefficients in D_{ij} can be named (D_{11}, D_{22}, D_{33}) or (D_X, D_Y, D_Z) or (D_L, D_T, D_T); they are in m^2/day.

Parameter $R \geq 1$ is the dimensionless *Retardation Coefficient*, due to adsorption of the solute on the solids (minerals): $R > 1$ if adsorption is present (retardation), else $R = 1$ (no retardation).

Coefficient λ is a decay coefficient $\lambda\left[\text{days}^{-1}\right]$, which represents a linear first order decay kinetic for the solute concentration (e.g., biodegradation). A complication occurs if both retardation and decay are present ($R > 1$ and $\lambda \neq 0$): in that case, it must be decided whether the decay kinetic is applicable not only to the solute but also to the adsorbed species: if yes, there is nothing to be changed in the PDE model (if not, then λ should be replaced by λ/R in the PDE).

Geometry, initial-boundary conditions, NAPL source model

The PDE problem of Eq. 6.1 will now be solved for specific initial and boundary conditions, involving also an auxiliary model of NAPL contaminant source. These conditions include several parameters in addition to the coefficients of the PDE Eq. 6.1.

The spatial domain for the PDE of Eq. 6.1 is the half-space $\Omega : \vec{x} \in \mathbb{R}^3 (x \geq 0)$. Furthermore, due to reflection symmetries with respect to the *xz-plane* and *xy-plane*, only the positive regions $y \geq 0$ and $z \geq 0$ need be calculated. In practice, therefore, only the 8th-fraction of space $x \geq 0$, $y \geq 0$, $z \geq 0$ need be considered. Still, in what follows, we still consider that the domain Ω is defined by $x \geq 0$.

- *Initial condition ($t = 0$):*

  The spatial distribution of initial concentration is prescribed throughout the domain:

  $$t = 0; \ C(\vec{x}, 0) = 0, \forall \vec{x} \in \Omega$$

- *Boundary conditions at infinity:*

  $$x \to +\infty; \ \lim_{x \to +\infty} C(x, y, z, t) = 0 \text{ and } \lim_{x \to +\infty} \frac{\partial C}{\partial x}(x, y, z, t) = 0 \ (\forall t \geq 0)$$

  $$y \to \pm\infty; \ \lim_{y \to \pm\infty} C(x, y, z, t) = 0 \text{ and } \lim_{y \to \pm\infty} \frac{\partial C}{\partial x}(x, y, z, t) = 0 \ (\forall t \geq 0)$$

  $$z \to \pm\infty; \ \lim_{z \to \pm\infty} C(x, y, z, t) = 0 \text{ and } \lim_{z \to \pm\infty} \frac{\partial C}{\partial x}(x, y, z, t) = 0 \ (\forall t \geq 0)$$

- *Boundary conditions in the vertical rectangular source patch at $x = 0$* (Fig. 6.1)

  A prescribed distribution of concentration C_0 in a rectangular part of the boundary plane $x = 0$ (see Fig. 6.1) serves as a way to model, indirectly, the presence of a trapped NAPL pollutant of given initial mass M_0. The effect of the NAPL is represented by imposing a fixed concentration C_0 within the rectangular patch, and this only during a finite duration t_S related to the loss of mass $M(t)$ of the NAPL source (this mass $M(t)$ is modeled separately in a simple manner, see below).

  Within the rectangular source patch (in the plane $x = 0$):

  $$x = 0; \ y \epsilon [-L_s/2; +L_s/2]; \ z \epsilon [-E_s/2; +E_s/2]$$

  $$\begin{cases} 0 < t < t_S : C(x, y, z, t) = C_0 \\ \quad t \geq t_S : \quad C(x, y, z, t) = 0 \end{cases} \tag{6.5}$$

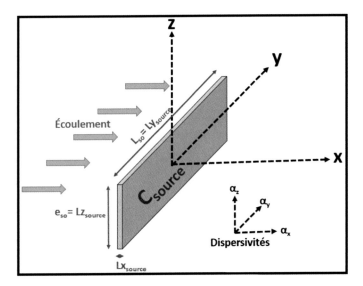

Fig. 6.1 Geometry of the DNAPL source in the 3D analytical model. The transport domain is the half-space $x \geq 0$. The rectangular source patch is located upstream, within the plane $x = 0$, having planar dimensions $Ly_{SOURCE} \times Lz_{SOURCE}$, and small width Lx_{SOURCE}. To emulate DNAPL dissolution, a concentration boundary condition $C = C_0$ [mg/l] is prescribed within the patch as long as DNAPL mass $M(t)$ remains positive ($t < t_S$). After $M(t)$ reaches zero, for times $t \geq t_S$, the condition within the patch becomes $C = 0$

Outside the rectangular source patch (in the plane $x = 0$):

$$x = 0; \ y \notin [-L_s/2; +L_s/2]; \ z \notin [-E_s/2; +E_s/2] :$$
$$C(x, y, z, t) = 0 \ (\forall t \geq 0) \tag{6.6}$$

Note, in the ESPER 1 code, the total width of the rectangular source along Oy is denoted $L_{Ysource}$ or L_{YS}, and its total height along Oz is denoted $L_{Zsource}$ or L_{ZS}. (Here we also use the notation or Es instead of Lzs for the height of the patch source).

- ***Model of decaying source mass $M(t)$ and source duration t_S***

The decaying mass of the NAPL source $M(t)$, and the duration of the source t_S, are obtained by solving separately a simplified auxiliary problem of coupled exchange, at equilibrium, between the DNAPL source and the advection of the dissolved contaminant (concentration C) through the source. The result of this auxiliary model is that the mass of the trapped DNAPL decreases linearly with time, until it vanishes at a finite time t_S (the duration of the NAPL source):

$$M(t) = M_0 - Q_0 \times t; \ with : Q_0 = C_0 E_s L_s V_0^* \theta_{kin} \tag{6.7}$$

…where C_0 is the imposed concentration in the rectangular source patch, (E_s, L_s) are vertical and horizontal side lengths of the rectangular patch, V_0^* is the groundwater pore velocity scaled by the retardation coefficient ($V_0^* = V_0/R_0$ or $V_0^* = V_X/R$), and θ_{kin} is kinematic porosity. The parameter Q_0 represents the rate of decay of the DNAPL mass: $\frac{dM}{dt} = -Q_0 [\text{kg/day}]$.

Now, the above model of decaying mass yields immediately the extinction time, or duration of the source t_S. It is obtained by letting $M(t_S) = 0$. This yields t_S as a function of the other parameters:

$$t_S = M_0 / \left(C_0 E_s L_s V_0^* \theta_{kin} \right) \tag{6.8}$$

Note that the DNAPL mass $M(t)$ starts by decreasing linearly, then reaches zero at $t = t_S$, and then remains zero thereafter for $t > t_S$. As a consequence, the concentration boundary condition on the source patch is $C = C_0$ for $0 < t < t_S$, then $C = 0$ for $t \geq t_S$ (see earlier boundary conditions).

6.1.2 The Quasi-analytical Solution C(x, y, z, t) and M(t)

The solution is constructed in several steps.

The first step is the solution $C^*(x, y, z, t - t')$ for an instantaneous patch source release at time $t = t'$, corresponding formally to a source concentration $C_0 \times \delta(t - t')$, where $\delta(t - t')$ is the Dirac delta "function" (distribution) centered at time $t = t'$.

$$C^*(x, y, z, t - t') = \frac{C_0 x}{8\sqrt{\pi D_X (t-t')^3}} \exp\left[\frac{V^2(t-t')}{4D_X} - \lambda(t - t') + \frac{V_x}{2D_X} - \frac{x^2}{4D_X(t-t')} \right]$$
$$\times \left\{ \text{erfc}\left[\frac{Y_1 - y}{2\sqrt{D_Y(t-t')}} \right] - \text{erfc}\left[\frac{Y_2 - y}{2\sqrt{D_Y(t-t')}} \right] \right\} \times \left\{ \text{erfc}\left[\frac{Z_1 - z}{2\sqrt{D_Z(t-t')}} \right] - \text{erfc}\left[\frac{Z_2 - z}{2\sqrt{D_Z(t-t')}} \right] \right\}$$
$$\tag{6.9}$$

…where the coordinates of the rectangular patch were denoted $[Y_1, Y_2]$ and $[Z_1, Z_2]$ for convenience. Thus, the patch source is defined by the rectangle $y \in [Y_1, Y_2]$ and $z \in [Z_1, Z_2]$. In our case, the patch is centered at the origin, and we have: $[Y_1, Y_2] = [\pm L_{Ysource}]$ and $[Z_1, Z_2] = [\pm L_{Zsource}]$.

The second step is to deduce the solution for a time continuous patch source of infinite duration, by integrating the previous instantaneous solution $C^*(x, y, z, t - t')$ from $t' = 0$ to $t' = t$, whence:

$$C(x, y, z, t) = \int_{\tau=0}^{\tau=t} C^*(x, y, z, \tau) d\tau \tag{6.10}$$

The third step consists in obtaining the solution for a finite duration NAPL source patch. Let "t_S" designate the finite duration of the source, which can be represented as a jump function, constant until $t = t_S$, then null for $t > t_S$. The jump can be represented by the classical Heaviside function $H(\tau)$ such that $H(\tau) = 0$ for $\tau < 0$ and $H(\tau) = 1$ for $\tau \geq 0$. The solution is obtained by convolution of $C^*(x, y, z, t - t')$ with $H(t_S - t')$ taking into account the finite duration t_S of the source:

$$C(x, y, z, t) = \int_{t'=0}^{t'=t} C^*(x, y, z, t - t') H(t_s - t') dt'$$

Taking into account the Heaviside function, this expression can be manipulated to obtain finally the required finite duration source solution C in terms of the instantaneous source solution C^*:

$$C(x, y, z, t) = \int_{\tau=Max(0,t-t_s)}^{\tau=t} C^*(x, y, z, \tau) d\tau \tag{6.11}$$

Finally, this solution can be adapted to take into account a retardation coefficient $R > 1$ (we have $R = 1$ in the previous solution). For the case $R > 1$, it suffices to rescale the groundwater velocity V and the dispersion coefficients D_{XX}, D_{YY}, D_{ZZ}, by dividing each of them by R. Since the dispersion coefficients incorporate both local isotropic diffusion D_0 and dispersion $\alpha_{XX} V_X, \alpha_{YY} V_X, \alpha_{ZZ} V_X$, the required scaling is: $\frac{D_0}{R} + \alpha_{XX} \frac{V_X}{R}, \frac{D_0}{R} + \alpha_{YY} \frac{V_X}{R}, \frac{D_0}{R} + \alpha_{ZZ} \frac{V_X}{R}$. Thus, to obtain the "retarded" solution, it suffices to let $D_{XX} = \frac{D_0}{R} + \alpha_{XX} \frac{V_X}{R}$ instead of $D_{XX} = D_0 + \alpha_{XX} V_X$, and similarly for D_{YY} and for D_{YY}, in the expression of the instantaneous solution C^* obtained at the first step above (6.9).

6.1.3 Visualizations of the Solution C(x, y, z, t) and M(t)

This section presents visualizations of the quasi-analytical solution for the concentration plume $C(x, y, z, t)$, and also the decaying NAPL source mass $M(t)$, for *deterministic* reference parameters.

First, recall that the 3D domain of the solution is the half-space $x \geq 0$, so the half-region $x < 0$ cannot be shown. Furthermore, only the region $y \geq 0$ needs to be shown due to reflection symmetry with respect to the XZ-plane, and only the region $z \geq 0$ needs to be shown due to reflection symmetry with respect to the XY-plane. Thus, in total, only an 8th-fraction of space is shown ($x \geq 0$, $y \geq 0$, $z \geq 0$).

The next few plots show deterministic snapshots of the advective–dispersive concentration plume $C(x, y, z, t)$ either as iso-concentration surfaces, or as axial concentration profiles, based on the semi-analytical solution presented in the previous section. A set of reference parameters is used, corresponding to the mean values of the uncertain parameters. The concentration in the source patch is $C_0 = 32$ mg/l, and the

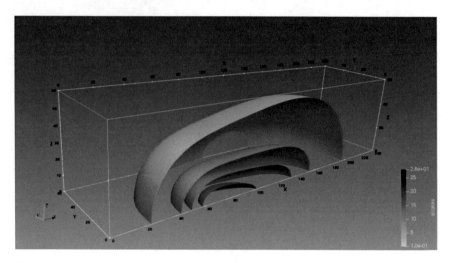

Fig. 6.2 Example solution from the 3D quasi-analytical model, showing a snapshot of the plume at time $t = 900$ days in the form of iso-concentration surfaces, from $C = 0.1$ mg/l (outer yellow surface) up to roughly 20 mg/l (inner dark green surface). The initial patch concentration is $C_0 = 32$ mg/l, and the source extinction time is $t_S = 600$ days. Parameters: $V = 0.5$ m/day; $(\alpha_{XX}, \alpha_{YY}, \alpha_{ZZ}) = \alpha = 1.25$ m; solute decay rate $\lambda = 0$; retardation coefficient $R = 3$

geometric parameters of the patch source are $Ys = Lo = 40$ m, $Zs = Eo = 20$ m. All spatial coordinates X, Y, Z are in meters, and times are in days.

- First, Fig. 6.2 shows a 3D snapshot of the plume $C(x, y, z, t)$ at time $t = 900$ days, in the form of iso-concentration surfaces. The source extinction time is $t_S = 600$ days, therefore the plume is shown here 300 days after extinction of the source. The parameters were chosen as follows: groundwater pore velocity $V = 0.5$ m/day; isotropic dispersivities $(\alpha_{XX}, \alpha_{YY}, \alpha_{ZZ}) = \alpha = 1.25$ m in all directions; first order decay of the solute was ignored ($\lambda = 0$); but solute adsorption was represented by a retardation coefficient $R = 3$.
- Secondly, Fig. 6.3 depicts axial concentration profiles at different times along the centerline ($y = 0, z = 0$). The selected times are comprised between 600 and 1200 days, while the source extinction time is $t_S = 600$ days. Therefore, all the axial profiles in this figure depict the axial evolution of concentration since the extinction of the NAPL source (times $t \geq 600$ days). The parameters are the same as in the previous 3D vizualization of Fig. 6.3.

6.1.4 Monte Carlo Analyses of 3D Analytical Model (ESPER1 Package)

This 3D model was implemented in the *ESPER-1 package*, which included some of the probabilistic uncertainty analyses developed within the ESPER project.

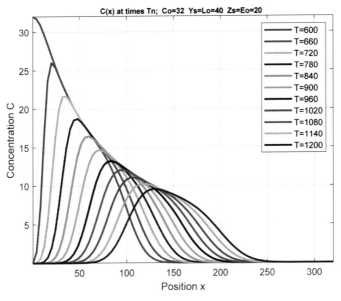

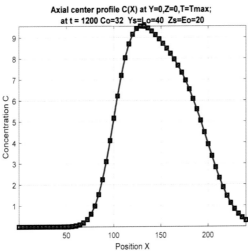

Fig. 6.3 Axial plots of the quasi-analytical 3D concentration plume for finite duration source. Axial concentration profiles $C(x, t)$ are shown on the centerline ($y = 0, z = 0$). **Top**: successive profiles $C(x, t_n)$ at different times t_n since the extinction of the source ($t \geq t_S = 600$ days), until time $t = 1200$ days. **Bottom**: zoom on the last concentration profile at time $t = 1200$ days. All input parameters are the same as in the previous 3D vizualization

ESPER-1 included, first, the 3D quasi-analytical model of concentration $C(x, y, z, t)$ and $Mass(t)$, and secondly, the probabilistic input/output Monte Carlo procedure to propagate uncertainty through this 3D model. More details on both aspects are provided in the ESPER public report by Chastanet et al. [1], describing this *ESPER-1 package* under the name *ESPER software v.1.0.*

Monte Carlo uncertainty analysis set up for the 3D quasi-analytical model

In this application, uncertainty analysis was developed by running direct Monte Carlo simulations on the model for probabilistic input parameters (random input parameters, treated one by one).

The empirical distribution of the output criterion (e.g., concentration at some point downstream from the source), was constructed directly from the multiple simulations via the quasi-analytical model (without a meta-model). A MATLAB script was programmed to calculate and plot this quasi-analytical solution $C(x, y, z, t)$ (Eqs. 6.9 and 6.11) on any 3D space–time grid, and to manage the multiple simulations required by the Monte Carlo procedure. The decaying mass $M(t)$ of the NAPL source, and its extinction time t_S, are also modeled as a simple but nonlinear function of the input parameters (Eqs. 6.7 and 6.8).

The uncertain inputs and the uncertain outputs, or criteria used for this problem are listed in Table 6.1.

Construction of the empirical concentration CDF: Proba($C(x1, t) < c$) versus time

Starting with multiple replicates of random input parameters, the Monte Carlo approach leads to multiple replicates of the model output. For instance, with $M = 200$ replicates of an input parameter like permeability or initial concentration, the 3D model at hand delivers $M = 200$ replicates of the 3D space–time concentration field $C(\vec{x}; t)$, which must then be analyzed probabilistically.

Figure 6.6 shows an example with the 3D semi-analytical transport model, where the uncertainty analysis focuses on the temporal evolution of the output concentration C at a given point $\vec{x_1}$. The empirical CDF of concentration $C(\vec{x}_1; t_n)$ is constructed at each different time t_n, using the M replicates of $C(\vec{x}_1; t_n)$, based on the CDF point estimation method of Hazen (Eq. 3.4).

In the case shown in Fig. 6.6, a single uncertain input parameter was considered, namely, the initial concentration C_0 in the rectangular source patch. About 200 Monte Carlo simulations of the 3D quasi-analytical model were executed ($M \approx 200$). The chosen target point was $(x_1, y_1, z_1) = (30\,m, 0, 0)$. As can be seen, the "postprocessing" of these Monte Carlo simulations yields the temporal evolution of the CDF of concentration at the target point versus time. By definition, the concentration CDF is the function $F_{C(t)}(c)$ defined as the probability of non-exceedance, that is:

$$F_{C(t)}(c) = Proba\{C(\vec{x_1}; t) < c\} \tag{6.12}$$

…where "c" is any given concentration level. The following type of information can be gathered from the results shown in Fig. 6.4: given a threshold concentration $c_{THRESHOLD} = 5$ (red vertical line), and a probability of non-exceedance $F = 0.90 = 90\%$ (blue horizontal line), one may obtain the time such that concentration "c" remains less than 5 mg/l with at least 90% probability beyond that time (that time is roughly 165 days from the figure).

Figure 6.5 shows a result similar to Fig. 6.4, with uncertain initial concentration in the source patch, but at a more remote target position, $(x_1, y_1, z_1) = (100\,m, 0, 0)$, and at larger time scales (up to 1200 days instead of 200 days). Also, a smaller number of Monte Carlo replicates were implemented here ($M \approx 50$ instead of 200 at each discrete time); for this reason, the Monte Carlo samples are now clearly visible in the concentration CDFs (circle symbols). Another difference in Fig. 6.5 compared with the previous figure is that, here, the time range is 600 days $< t <$ 1200 days, where $t_S = 600$ days is the time of total extinction of the NAPL source (zero mass). In other words, in Fig. 6.5, we are looking at the uncertainty of the migrating advective–dispersive concentration plume *after* the total extinction of the NAPL source (the NAPL source mass has vanished and is null at times $t \geq 600$ days).

Table 6.1 Uncertain input parameters and output criteria for the 3D semi-analytical model

Input parameters	Notation	Units	Probability distribution
Initial concentration	C_0	mg/l	Uniform $\sim [16, 48]$ mg/l Mean:~ 32 mg/l Coeff. of variation: $\sim 100\%$
Groundwater velocity	V	m/day	Uniform or Log-normal Mean: 1 m/day typically
Initial mass of NAPL source	M_0	kg	Uniform or Log-normal Mean: 320 kg typically
Output criteria			
Concentration at target point $\vec{x_1}$, such as $\vec{x_1} = (x_1, 0, 0)$, downstream on the centerline	$C(\vec{x_1}, t)$	mg/l	Empirical probability law of $C(\vec{x_1}, t)$: $F_{C(t)}(c) = Proba\{C(\vec{x_1}; t) < c\}$
Mass of the NAPL source	$M(t)$	kg	Empirical probability law of $M(t)$: $F_{M(t)}(m) = Proba\{M(t) < m\}$
Duration of the NAPL source (extinction time t_S)	t_S	days	Empirical probability law of t_S: $F_{t_S}(\tau) = Proba\{t_S < \tau\}$

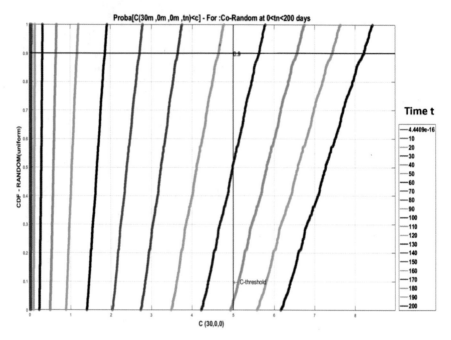

Fig. 6.4 Temporal evolution of empirical CDF of concentration, "$F_{C(t)}(c)$", at a given target point $\vec{x}_1 = (30\,\text{m}, 0, 0)$. Uncertainty on $C(\vec{x}_1, t)$ is due to random initial concentration C_0 of the DNAPL patch source (C_0 is uniformly distributed). Uncertainty was propagated via $M \approx 200$ Monte Carlo runs of the model. The CDF, $F_{C(t)}(c) \equiv Proba\{C(\vec{x}_1, t_n) \leq c\}$, is shown at discrete times $t_n = 0, 10, 20, \ldots, 190, 200$ days. The abscissa is concentration "c", ranging from $c = 0$ to $c = 9$ mg/l

6.2 Uncertainty Analysis of 3D Solute Migration and Source Attenuation in a Polluted Site (MODFLOW-SURFACT)

We present here finally an application of some previously reviewed techniques of uncertainty analyses to a 3D problem of advection–dispersion coupled with DNAPL source attenuation in an actual polluted site, modeled with the MODFLOW-SURFACT™ computer code. We have focused on HVOC components (Halogenated Volatile Organic Compounds) released by the DNAPL source, like PCE, TCE, DCE, etc. We have then developed uncertainty analyses using the *meta-model approach* implemented with the OpenTURNS software. For a brief initial account of these field site analyses, see ([7], in French). Concerning OpenTURNS, see [8].

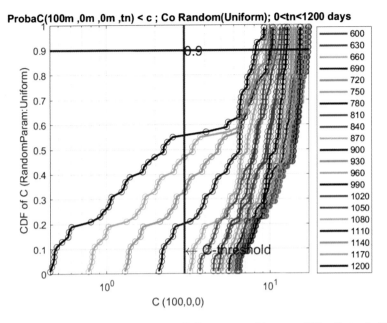

Fig. 6.5 Temporal evolution of concentration CDF at target position $x_1 = 100$ m (instead of 30 m in Fig. 6.6), and larger time scale (up to 1200 days instead of 200 days in Fig. 6.6). The time range here is 600 days $< t <$ 1200 days, where $t_S = 600$ days marks the total extinction of the NAPL source. Compared to Fig. 6.6, a smaller number of Monte Carlo replicates were used: $M \approx 50$ at each discrete time (instead of $M \approx 200$). Monte Carlo samples are fewer on each CDF curve, so they are more clearly visible (circle symbols). The abscissa shows log-concentrations ($\log_{10} c$), with c in the range ~ 0.5 to 15 mg/l

6.2.1 The Modflow-Surfact Code and Model Equations

The commercial code MODFLOW-SURFACT™ (version 2011) is also named MODHMS/MODFLOW-SURFACT (Hydrogeologic Inc.). The code and the model equations used here are briefly described below.

MODHMS™ is an integrated surface/groundwater flow code developed by HydroGeoLogic Inc., and MODFLOW-SURFACT™ is the groundwater flow module of MODHMS™, based on enhancements of the U.S. Geological Survey modular 3D groundwater flow code MODFLOW [9, 10], which later evolved into MODFLOW-2000 [11]. MODFLOW-SURFACT™ is written in FORTRAN. It simulates subsurface flow and contaminant transport. It can accommodate 3D subsurface flow based on Darcy's law, and various modalities of species transport. It can solve 3D saturated/unsaturated flow equations or alternatively, saturated flow equations for unconfined aquifers (it can also include both confined and unconfined layers).

Concerning subsurface transport, MODFLOW-SURFACT™ has the following capabilities (*some but not all of these were used in this study*): NAPL (Non-Aqueous Phase Liquid) source release in groundwater and in the unsaturated zone; vapor flow

Fig. 6.6 Plane view of the DCE concentration plume (*DiChloroEthene*) obtained with the post-excavation calibrated parameters at the polluted field site, simulated with the MODFLOW-SURFACT™ computer code

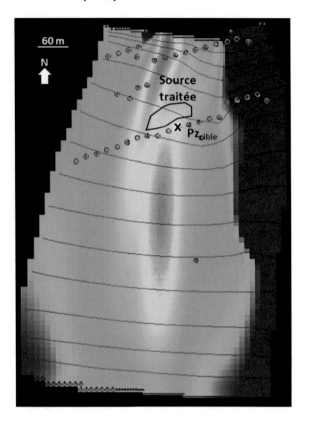

extraction and air sparging; multi-phase and multi-component contaminant transport with biodegradation; and (in the *Reaction Module*) various reactions of mobile and immobile chemical species for dissolution or biodegradation. The flow-transport system can be density-dependent.

Numerically, MODFLOW-SURFACT™ is based on block-centered finite-difference discretization in space (for the flow solver), and it contains an iterative Newton–Raphson package for linearization, a Preconditioned Conjugate Gradient (PCG) package for solving linear systems, a special adaptive Total Variation Diminishing scheme for the transport solver, and various options for adaptive time-stepping.

In this study, the MODFLOW-SURFACT™ computer code was implemented for simulating subsurface flow and coupled contaminant transport in a 3D domain with several curvilinear layers. The full 3D equational model is complex; it includes a broad set of phenomena both in the NAPL pollution source, and in the resulting concentration plume, such as: NAPL dissolution in groundwater, adsorption/desorption mechanisms, advective–dispersive transport, and a biodegradation sequence specific to chloro-ethenes. This study is part of a broader framework on depollution and remediation of French sites polluted by Dense Non Aqueous Phase

Liquids (DNAPL) and by the subsequent migration of Halogenated Volatile Organic Compounds (HVOCs).

For convenience, let us present here a simplified version of the 3D model, to illustrate the flow and transport properties and the NAPL dissolution mechanism to be solved by MODFLOW-SURFACT™:

$$\begin{cases} C\partial H/\partial t = div\left\{K\overrightarrow{grad}H\right\} \ and \ \vec{V} = -K(\vec{x})\overrightarrow{grad}H \\ \frac{\partial}{\partial t}\{\Phi(1-S)C\} + div\left\{\vec{V}C\right\} - div\left\{\Phi(1-S)\underline{\underline{D}}\overrightarrow{grad}C\right\} = \\ \qquad -\alpha_*\Phi(1-S)\{C - C_{EQ}\} \\ \frac{\partial}{\partial t}\{\rho_{NAPL}\Phi S\} = +\alpha_* \Phi (1-S)\{C - C_{EQ}\} - \lambda\,\rho_{NAPL}\Phi S \end{cases} \tag{6.13}$$

This equational system, to be solved numerically with MODFLOW-SURFACT™, governs solute concentration $C(x, y, z, t)$, coupled to trapped "oil" saturation $S(x, y, z, t)$. In this study, we refer to the steady state flow situation, with steady hydraulic head $H(\vec{x})$ and velocity field $\vec{V}(\vec{x})$. NAPL dissolution and solute migration are transient processes. The migrating concentration plume $C(\vec{x}, t)$ is due to dissolution of the trapped DNAPL (Dense Non Aqueous Phase Liquid), which is the contaminant "source". Concentration is advected by the Darcy groundwater velocity field $\vec{V}$ or $V_i(i = 1, 2, 3)$ [m/day], and is dispersed according to the anisotropic dispersion tensor $D_{ij}(i, j = 1, 2, 3)$ [m²/day].

The 2nd line of Eq. 6.13 is a balance equation for solute mass per m³ of porous medium (it is the dissolved contaminant present in the liquid phase), and the 3rd line of Eq. 6.13 is a balance equation for the trapped DNAPL mass per m³ of porous medium. The dissolution model is assumed linear, with constant mass exchange coefficient α_* (*in fact α_* should decrease with saturation, and the model is improved by letting $\alpha_* = 0$ when saturation reaches some low residual value $S_{RESIDUAL}$*). The exchange model can be reformulated by introducing another mass exchange coefficient defined as $\alpha = \alpha_*\Phi(1 - S)$ instead of α_*. Both α_* and α have units of [day^{-1}]. Overall, the variables and parameters in Eq. 6.13 are similar to those defined previously in the 1D transport model of Eq. 5.2.

To sum up, the variables of Eq. 6.13 are defined as follows:

• $C(\vec{x}, t)$: solute concentration, $\frac{kg}{m^3 water}$ or $\frac{g}{m^3 water}$ (NAPL constituent dissolved in the water phase)
• $S(\vec{x}, t)$: NAPL phase saturation (also named S_O for "oil saturation") in $\left[\frac{m^3 NAPL}{m^3 Pores}\right]$

...with either deterministic or uncertain parameters, defined as follows:

• Φ : aquifer porosity in m³/m³ (also denoted ε)
• $\vec{V}$: Darcy velocity vector $V_i(i = 1, 2, 3)$ in m/day (*NB: pore water velocity is $\vec{V}_{PORE} = \vec{V}/\Phi$*)
• $\underline{\underline{D}}$: dispersion coefficient D_{ij} ($i = 1, 2, 3$; $j = 1, 2, 3$) in m²/day; it can be interpreted as local isotropic diffusion + tensorial dispersion; its simplest form in the reference frame aligned with horizontal velocity $\vec{V} = (V_1, 0, 0)$ is a diagonal matrix given by: $D_{11} = D_0 + a_{11}.|V_1|$, $D_{22} = D_0 + a_{22}.|V_1|$, $D_{33} = D_0 + a_{33}.|V_1|$,

where $a_{11} = a_L$[m] is the longitudinal dispersivity (horizontal), and $a_{22} = a_{33} = a_T$[m] are transverse dispersivities (horizontal and vertical).

- ρ_{NAPL}: density of the trapped NAPL, in $\left[\text{kg } NAPL/\text{m}^3 \ NAPL \right]$.
- C_{EQ}: equilibrium concentration of the solute (concentration at which NAPL dissolution stops)
- α_*[day^{-1}]: Mass exchange coefficient for NAPL dissolution, in $\left[\left(\frac{\text{kg } Solute}{\text{m}^3 \ domain} \right) / \left(\frac{\text{kg } NAPL}{\text{m}^3 \ domain} \right) \right]$ per day. Another version of this coefficient can also be defined as follows: $\alpha = \alpha_* \Phi (1 - S)$, which has the same units, [day^{-1}].
- λ: biodegradation decay parameter [day^{-1}] of the NAPL pollutant (λ^{-1} is related to half-life).

6.2.2 Main Hydrogeologic Characteristics of the Polluted Test Site

The polluted test site is a 9 ha maintenance center for vehicles, which has been closed for decades. It is polluted with chlorinated solvents. It is located over a loam and gravel aquifer, resting on a marn substratum at 7–8 m depth. There is a stream nearby. During the years 2010–2020, several DNAPL sources were identified at depth, and they were monitored. This study focuses on one DNAPL source which was distributed vertically in a rather heterogeneous fashion down to the substratum (total mass ~ 1.5 ton). Excavation works, aiming at remediation of the site, reduced the pollution by about 80%. The pollution that remained was present essentially in a sorbed state.

As a preliminary procedure, before uncertainty analyses, the MODFLOW-SURFACT™ model described in the previous subsection was calibrated on the pre-excavation and post-excavation states. The DCE concentration plume (*DiChloroEthene*) obtained with the post-excavation calibrated parameters is shown in Fig. 6.6.

6.2.3 Uncertainty Analysis: Metamodeling Procedure for the Field Site

In this study of uncertainty propagation, the *metamodel approach* will be used. As was seen in previous sections (Sects. 2.2.2 and 5.4.1), this approach requires constructing a response function for a pre-defined output criterion. The response criterion could be the time it takes for the NAPL source to decrease to less than 5% of its initial mass. The response criterion chosen in this study is the time (t_{CRIT}) taken for DCE concentration to reach an acceptable target value less than or at most equal to 400 µg/l at a target piezometer located immediately downstream from the NAPL source. This "critical time" criterion is therefore such that:

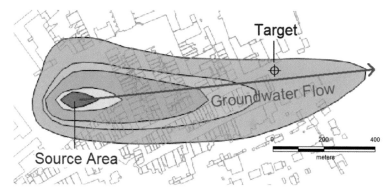

Fig. 6.7 Schematic illustrating the relative positions of the NAPL source zone, the solute concentration plume, the mean flow direction (left to right), and the downstream "target piezometer" which is used to define the critical time criterion. This target piezometer can be a pumping well

$$C(x_{TARGET}, t_{CRIT}) \leq C_{TARGET} = 400\,\mu g/l \qquad (6.14)$$

Empirically, the target concentration of 400 μg per liter is the concentration that can be expected after the treated NAPL source has been completely dissolved. The typical value of critical time is $t_{CRIT} \approx 6$ years based on the post-excavation calibrated model (not based on uncertainty analysis). The typical position of the target piezometer is illustrated in Fig. 6.7.

To sum up, we are interested in assessing the uncertainty of the critical time t_{CRIT} defined by Eq. 6.14. For this purpose, the uncertain input parameters must first be selected. After a preliminary screening of the most sensitive parameters based on a number of simulations with MODFLOW-SURFACT™, several uncertain parameters were retained, along with some output criteria to be analyzed, as shown in Table 6.2. Note: several of the uncertain input parameters that have been considered at different steps of the study are listed in the table, but detailed results are only presented for the first, hydraulic conductivity.

Then, the following steps were then implemented for a systematic uncertainty analysis of model results, based essentially on the *metamodel* approach (reviewed earlier in this work):

- **Step 1. Defining the range of variation of the uncertain parameters.**

  Note: this step could also involve defining the cross-correlations between the different parameters, although in this work, we did not consider such cross-correlations.

- **Step 2. Creation of a sampling plan for generating multiple replicates of input parameters**

  This step consists of selecting a limited number of replicates of the parameter set, judiciously distributed in the space of uncertain parameters (e.g., not necessarily uniformly but proportionally to their probability of occurrence). A simulation of

Table 6.2 Some uncertain input parameters and output criteria used for the 3D field site model in MODFLOW-SURFACT™ (details are given only for the first parameter, hydraulic conductivity)

Input parameters	Reference value	Probability law	Moments
Hydraulic conductivity $K\,[\text{m/day}]$ or $[\text{m/s}]$	$K = 1.1\ 10^{-4}\,\text{m/s}$	Log-normal	Mean: $m_K = 1.1\ 10^{-4}\,\text{m/s}$ Standard deviation: $\sigma_K = 1.6\ 10^{-4}\,\text{m/s}$
NAPL biodegradation decay parameter $\lambda\,[\text{day}^{-1}]$ or half-life $\tau_{50\%}\,[\text{days}]$	$\tau_{50\%}(\text{TCE}) \sim 100$ days, $\tau_{50\%}(\text{DCE}) \sim 200$ days		
Dispersivities, longitudinal and transverse: $a_L\,[\text{m}]$ and $a_T\,[\text{m}]$	Range: $a_L \in [0, 50\,\text{m}]$ Choice: $a_T \approx a_L/3$		
Initial NAPL saturation $S_{NAPL}(0)$ and/or initial mass of NAPL source $M(0)\,[\text{kg}]$			
Output criteria			
Critical time t_{CRIT} (Eq. 6.14)	$t_{CRIT} \sim 6$ years	(See probabilistic results on t_{CRIT})	

the 3D numerical model is then launched for each replicate of the parameter set. The «response» of each of these simulations is then collected and represented by points in the space of the uncertain parameters.

Observe that the output «response» is calculated only for the predefined criterion (it must be recalculated if the criterion is changed). This criterion could be some measure of "source lifetime", like the time it takes for the NAPL mass $M(t)$ to decrease to 5% of its initial mass $M(0)$, or it could be the previously defined critical time t_{CRIT} taken to reach a target concentration. We focus on the latter criterion.

Here, $M = 30$ replicates were selected for the multiparameter set, each replicate comprising 5 input parameters. They were sampled optimally in the 5-dimensional space of the parameters according to the "quasi random" sampling plan available in the Design of Experiments packages of OpenTURNS (see sampling plan shown previously in Sect. 5.4.1, Fig. 5.4, for the 1D model with 2 parameters). The response of the 3D numerical model was calculated point by point for each replicate, and finally, a 5-dimensional second order polynomial *metamodel* was fitted to the 30 points. The results indicate that the uncertain response criterion (t_{CRIT}) was distributed between -4 years and $+14$ years. The negative values are obviously an artefact due to the approximate nature of the metamodel. In spite of this, the nonlinear regression correlation "R" was fair ($R \approx 0.94$). The corresponding root-mean-square error (scaled) is evaluated as

$\sqrt{1 - R^2} \approx 0.24 = 24\%$. This implies that the fitted polynomial response (the metamodel) explains about 76% of the "true" response of the computer model.

- **Step 3. Construction of a metamodel.**

From the sample points selected in the previous step, as stated above, we have constructed a continuous response function (response surface) which passes near the points (polynomial best fit). This construction required running the MODFLOW-SURFACT™ model a number of times for different values of the set of input parameters. But once the response surface is described analytically, it constitutes the metamodel, which can now be used repeatedly for Monte Carlo analyses instead of the original model. That is, the analytical response function is used for Monte Carlo uncertainty analyses instead of complex and costly MODFLOW-SURFACT™ simulations.

In this application, note that we considered only a few uncertain parameters (rather than, say, several hundreds of uncertain parameters). We chose a simple nonlinear polynomial regression to obtain the response function from a moderate number of sample points (on the order of one hundred typically), where each sample point corresponds to running the complete model (MODFLOW-SURFACT™) for a given replicate of the set of uncertain parameters. Note also that our uncertainty analysis is a priori multivariate: two or more uncertain input parameters were considered simultaneously when plotting the resulting response function (at least in the initial phase of this study).

- **Step 4. Sensitivity analysis of model outputs with respect to input parameters.**

First, recall that sensitivity analysis is an integral part of Uncertainty Quantification. We have noted earlier that propagating uncertainty from inputs to outputs in a model can serve sensitivity analysis purposes, for instance by taking the uncertain inputs one-by-one (Sect. 2.3.1). In fact, sensitivity analysis is a topic in itself. The book by Da Veiga et al. [12] treats various approaches to uncertainty quantification via the «R» project, with a focus on Sensitivity Analyses in their Chaps. 5 and 6 based on Sobol Indices. These are not detailed in the present work, except briefly in Sects. 2.2.2 and 2.2.3. The reader is referred to [13] concerning Sobol indices and their particular relation to Polynomial Chaos metamodels.

Here, in the 3D field site study, we have calculated indicators to quantify the sensitivity of the metamodel response with respect to the uncertain parameters taken one-by-one (instead of simultaneously). Based on the constructed *metamodel*, several types of indices (such as Sobol indices) were calculated in order to assess model sensitivity with respect to input parameters. The results indicate that the most sensitive parameters were (1) hydraulic conductivity K, and (2) NAPL (TCE) biodegradation half-life (or decay parameter λ). However, overall, the sensitivities of other parameters were on the same order, and no clear hierarchy could be inferred.

- **Step 5. Uncertainty propagation via Monte Carlo simulations through the meta-model.**

This step consists in applying the Monte Carlo technique to the meta-model. It consists here in generating multiple replicates of the probabilistic inputs (sets of parameters with given probability distributions), and then, obtaining for each replicate the response given by the pre-constructed metamodel. The metamodel itself is not costly to implement: it is therefore applied to a large number of replicates. The response of the metamodel is then characterized in probabilistic terms; for instance the "source lifetime", or else the critical time t_{CRIT} corresponding to a target concentration, is analyzed in terms of its mean, its standard deviation, its Probability Density Function (PDF), and most importantly, its Cumulated Distribution Function (CDF).

In this last step, the metamodel serves as a surrogate for the full numerical model. It should be emphasized that the metamodel furnishes a response that depends on the proposed criterion to be analyzed (such as t_{CRIT}). The metamodel needs to be entirely reconstructed if a new output criterion is defined (there is a different metamodel for each different output criterion).

6.2.4 Results: Probabilistic Characterization of Output Criterion (Critical Time)

The Monte Carlo simulations to be performed on the metamodel could be carried out in principle with all 5 input parameters jointly. However, given the results of preliminary sensitivity analyses, it was decided to propagate uncertainty by considering solely the uncertainty of hydraulic conductivity K, which was considered lognormally distributed, and is very variable, with a coefficient of variation of 145% (see Table 6.2). The Monte Carlo simulations, performed on the metamodel, finally lead to probabilistic characterization of a single uncertain variable: the output criterion t_{CRIT} (Eq. 6.14).

Thus, an empirical PDF histogram and a pointwise CDF are estimated for t_{CRIT}, using the probability estimators explained earlier in Sect. 3.1.1. The resulting PDF and CDF of t_{CRIT} are shown in Fig. 6.8. The CDF of the response criterion (t_{CRIT}) can be used to make probabilistic statements concerning the uncertain response. Thus, a probabilistic answer can be given to a question like this:

What is the critical time $t_{CRIT95\%}$ which has a 95% probability of not being exceeded?

$$Proba\{t_{CRIT} \leq t_{CRIT\,95\%}\} = 0.95 \rightarrow find\ t_{CRIT\,95\%} \qquad (6.15)$$

Graphically, the empirical CDF of Fig. 6.8 shows that the 95% critical time is about 6.5 years:

$$Proba\{t_{CRIT} \leq t_{CRIT95\%}\} = 0.95 \rightarrow t_{CRIT95\%} \approx 6.5\ years$$

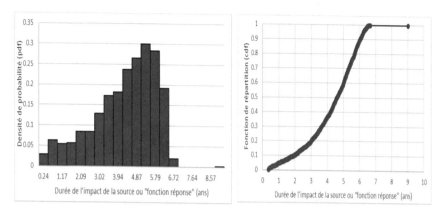

Fig. 6.8 Empirical probability distribution of the response criterion t_{CRIT} obtained from the 3D computer code via metamodeling: Probability Density Function histogram (at left) and Cumulated Distribution Function (at right). The uncertain input parameter (hydraulic conductivity K) is lognormally distributed and has a large coefficient of variation (145%)

It can also be seen that there is an 80% probability that the critical time be comprised between 2 and 6 years. The median value of the critical time, having 50% chances of being exceeded, is:

$$t_{CRIT}^{MEDIAN} \approx 4.5 \text{ years}$$

Finally, the most frequent value of t_{CRIT} can be obtained roughly from the mode of the PDF histogram:

$$t_{CRIT}^{MODE} \approx 5 \text{ years}$$

References

1. J. Chastanet, J.-M. Côme, R. Ababou, M. Quintard, M. Marcoux, N. Tribouillard, *Project ESPER—Evaluation of the sensitvity of prediction models for NAPL sources extinction and remediation: deterministic and probabilistic approaches to secure management decision (ESPER software Version 1.0—User's Guide)*. Public Report, ADEME, France, May 2019, 26 pp. [in English] (2019)
2. E.J. Wexler, Analytical solutions for one-, two-, and three-dimensional solute transport in ground-water systems with uniform flow. *USGS Report TWRI-3-B7*, pp. 1–190, in *Techniques of Water Resources Investigations of the US Geological Survey, Chap.B-7, Book 3, Applications of Hydraulics*. Washington DC (1992)
3. B. Sagar, Dispersion in three dimensions: approximate analytical solutions. ASCE J. Hydraul. Div. **108**(HY1), 47–62 (1982)
4. P.A. Domenico, An analytical model for multidimensional transport of a decaying contaminant species. J. Hydrol. **91**, 49–58 (1987)

5. M.R. West, B.H. Kueper, M.J. Ungs, On the use and error of approximation in the Domenico (1987) solution. *Ground Water* 2007 March–April; **45**(2), 126–135.
6. J.-S. Chen, L.Y. Li, K.-H. Lai, C.-P. Liang, Analytical model for advective-dispersive transport involving flexible boundary inputs, initial distributions and zero-order productions. J. Hydrol. **554** (2017). https://doi.org/10.1016/j.jhydrol.2017.08.050
7. J. Chastanet, M. Quintard, R. Ababou, J.-M. Côme, M. Marcoux, N. Tribouillard, «Prédire la durée d'impact d'une source de pollution: comment quantifier les incertitudes ?». *Quatrièmes RNRSSP: Rencontres Nationales Recherche Sites & Sols Pollués, 26–27 Nov. 2019, Le Beffroi de MontRouge - Portes de Paris, France.* Communication with extended abstract in French (5 pp.) (2019)
8. M. Baudin, A. Dutfoy, B. Iooss, A.-L. Popelin, OpenTURNS: An industrial software for uncertainty quantification in simulation, in *Handbook of Uncertainty Quantification*, ed. by R. Ghanem, D. Higdon, H. Owhadi (Springer, 2017). HAL-01107849v2 (46 p.)
9. M.G. McDonald, A.W. Harbaugh, *A Modular Three Dimensional Finite-Difference Ground-water Flow Model.* U. S. Dept. Interior and U. S. Geol. Survey, Nat. Center, Reston, Virginia, 528 pp. (1984)
10. M.G. McDonald, A.W. Harbaugh, *A modular three-dimensional finite-difference groundwater flow model.* U.S. Geol. Survey, Techniques of Water-Resources Investigations, Book 6, Chap. A1, 548 pp. (1988)
11. A. Harbaugh, E. Banta, M. Hill, M. McDonald, *Modflow-2000, the U.S. Geological Survey modular ground-water model—user guide to modularization concepts and the ground-water flow process.* U.S. Geological Survey, Open-File Report 00-92, 121 p. (2000)
12. S. Da Veiga, F. Gamboa, B. Iooss, C. Prieur, Basics and trends in sensitivity analysis (Theory and Practice in "R"). *SIAM—Society for Industrial and Applied Mathematics*, Philadelphia PA, xvi+291 p. (2021). https://doi.org/10.1137/1.9781611976694. https://epubs.siam.org/doi/abs/10.1137/1.9781611976694
13. L. Le Gratiet, S. Marelli, B. Sudret, Metamodel-based sensitivity analysis: Polynomial chaos expansions and Gaussian processes, in *Handbook of Uncertainty Quantification*, ed. by R. Ghanem, D. Higdon, H. Owhadi (2016). https://doi.org/10.48550/arXiv.1606.04273

Chapter 7
Discussion and Conclusions

In this book, we have presented several concepts and methods for propagating uncertainty through models, especially environmental geoscience models and hydrogeological models. One of the examples involves a probabilistic model of corrosion pit growth on nuclear waste canister, but most other examples involve solute concentration migration or decay: 0-dimensional first order decay kinetics, 1D concentration transport, and fully 3D concentration transport models (analytical and computational) in the presence of a trapped NAPL "source". The novel 3D quasi-analytical model of advective–dispersive contaminant migration from a decaying rectangular patch of NAPL source may be of interest not only for uncertainty analyses, as we do here, but also for benchmark tests in contaminant hydrogeology. The final sub-section dealing with the 3D numerical modeling of a real polluted site with the MODFLOW-SURFACT code illustrates how uncertainty analysis methods can be employed in practice, with what objectives, what output criteria, and what kinds of results.

A broad range of uncertainty propagation methods is covered in the theoretical sections, and then illustrated with the example models. Conceptually, the methods investigated here are based on probabilistic concepts, on fuzzy variable concepts, or both in the case of the "possibilistic" approach. Technically, our discussion covers several aspects: sampling plans such as LHS and other plans ("Design of Experiments"); different ways of implementing Monte Carlo simulations (directly on the model itself, or indirectly via a "metamodel" or "surrogate model"); various ways of propagating uncertainty analytically (exactly or approximately through Taylor expansions); and other issues of interest like multivariate probabilities to deal with joint sets of uncertain parameters, and sensitivity analyses performed via uncertainty propagation.

R. Ababou et al., *Uncertainty Analyses in Environmental Sciences and Hydrogeology*,
SpringerBriefs in Applied Sciences and Technology,
https://doi.org/10.1007/978-981-99-6241-9_7

Printed in the United States
by Baker & Taylor Publisher Services